LESS THAN HUMAN

ALSO BY DAVID LIVINGSTONE SMITH

Why We Lie
The Most Dangerous Animal

LESS THAN HUMAN

Why We Demean, Enslave, and Exterminate Others

David Livingstone Smith

ST. MARTIN'S PRESS NEW YORK

For my aunt, Laura Forester Mueller

LESS THAN HUMAN. Copyright © 2011 by David Livingstone Smith. All rights reserved. Printed in the United States of America. For information, address St. Martin's Press, 175 Fifth Avenue, New York, N.Y. 10010.

www.stmartins.com

Library of Congress Cataloging-in-Publication Data

Smith, David Livingstone, 1953–
    Less than human : why we demean, enslave, and exterminate others / David Livingstone Smith.—1st ed.
        p. cm.
    ISBN 978-0-312-53272-7
    1. Humanity—Psychological aspects.   2. Social isolation.
3. Marginality, Social.   4. Social status.   I. Title.
    HM1131.S65 2011
    305.5'68—dc22

                            2010040196

First Edition: March 2011

10 9 8 7 6 5 4 3 2 1

CONTENTS

ACKNOWLEDGMENTS

THIS BOOK WAS DIFFICULT TO WRITE, and I couldn't have completed it without the help of many people. I owe most to my wife, Subrena Smith, whose incisive comments and exacting intellectual standards kept bouts of intellectual enthusiasm from descending into sloppy reasoning. She has helped me avoid many errors and oversights, and her unflagging concern for my well-being gave me the mental and physical space to concentrate on research and writing.

I wouldn't have considered undertaking this project were it not for the urgings of my friend and colleague Anouar Majid, who insisted that I should devote my energies to investigating dehumanization at a time when I was considering various possible projects, and was uncertain which one to choose. "You've got to do it, David!" he said to me enthusiastically over dinner one evening. "Everyone talks about dehumanization, but hardly anyone theorizes it." As I soon discovered, he was absolutely right.

The book that you are reading now would have been nowhere near as readable had it not been for the efforts of Bruce Schneier, who carefully scrutinized and insightfully commented on the penultimate (or rather pen-penultimate) draft of the manuscript.

I would also like to give heartfelt thanks to my literary agent Michael Psaltis, and my editor at St. Martin's Press, Daniela Rapp, who displayed superhuman patience accommodating my trail of broken promises ("I'll get the manuscript to you in three months, for *certain*"). Finally, I'd like to thank all the library staff at the University of New

England, who promptly and efficiently responded to my seemingly interminable requests for books and journal articles.

In addition to these individuals, there are many others who have contributed to the writing of this book by generously sharing their ideas, clarifying facts, and directing me to sources of information, including Zadie Beagle, Vahan Dadian, Raymond Evans, Daniel Gilbert, Chris Hedges, Karl Jacoby, Scott MacDonald, David Mandel, Laura Mueller, Paul Roscoe, Michael Sargent, Wolfgang Wagner, Adam Waytz, Charlotte Witt, and Richard Wrangham. Thanks to you all.

I understand
what you want your filthy slave to be. I am
your barbarian, your terrorist;
your monster.

—ALI ALIZADEH, "YOUR TERRORIST"

PREFACE

CREATURES OF A KIND SOMEWHAT INFERIOR

We hold these truths to be self-evident, that all men are
created equal, that they are endowed by their Creator
with certain unalienable Rights, that among these are
Life, Liberty and the pursuit of Happiness.

— THOMAS JEFFERSON,
UNITED STATES DECLARATION OF INDEPENDENCE

THESE THIRTY-FIVE WORDS ARE OFTEN QUOTED reverently. The
ideal that they express—the principle that all men (that is, all human
beings) have certain basic rights *just because they are human*—is easy to
resonate with, and to applaud. But Jefferson's words beg a vexing question: the question of who, exactly, should be counted as human.

Jefferson's contemporaries weren't certain.[1] The uneasy relationship
between the economic attractions of slavery and the Enlightenment
vision of human dignity was a long-standing one, and for those torn
between the demands of conscience and the seductions of self-interest,
there was a way out of the dilemma. They could deny that African
slaves were human, and in this way they could square the moral circle.
By dint of a sleight of mind, the very men who insisted on the God-given
right of all humankind to liberty could, in good faith, countenance
and participate in the brutal and degrading institution of slavery. Many
of the great thinkers of the Enlightenment, who had championed the
concept of individual human rights and defined the philosophical

1

underpinnings of the great American experiment, routinely excluded nonwhites from the category of the human. The idea that the towering figures of the eighteenth century were champions of liberty for all is, in the words of Northwestern University philosopher Charles Mills, "profoundly misleading, deeply wrong." It "radically mystifies the recent past, and . . . needs to be confronted and discredited if our sociopolitical categories are to be true to the world that they are supposed to be mapping."[2]

It wasn't just the highbrows who thought of blacks as less than human. The theoretical views of the intellectuals—the philosophers, statesmen, and politicians—merged seamlessly with ideological beliefs which, however poorly articulated, had long been entrenched in the popular consciousness. Thus, the pseudonymous author of *Two Dialogues on the Man-Trade*, an abolitionist tract published in 1760, observed that:

> [T]hose who are concerned in the Man Trade have . . . a confused imagination, or half formed thought, in their minds, that the Blacks are hardly of the same Species with the white Men, but are Creatures of a Kind somewhat inferior . . .

"I do not know how to think that any white Men could find it in their Hearts," the author continued, "that the common Sentiments of Humanity would permit them to treat the black Men in that cruel, barbarous Manner in which they do treat them, did they think and consider that these have rational, immortal Souls."[3] Subhumans, it was believed, are beings that lack that special something that makes us human. Because of this deficit, they don't command the respect that *we*, the truly human beings, are obliged to grant one another. They can be enslaved, tortured, or even exterminated—treated in ways in which we could not bring ourselves to treat those whom we regard as members of our own kind.

This phenomenon is called *dehumanization*. It is the subject of this book.

Before I began to investigate dehumanization, I assumed that there was substantial research literature devoted to it. The fact that dehumanization is mentioned so frequently, both in popular journalism and in academic writings, led me to believe (wrongly, it turned out) that it had already been extensively studied. Then, as I began to hunt for writings on dehumanization, it dawned on me that although scholars from a wide range of disciplines are convinced that dehumanization plays a crucial role in war, genocide, and other forms of brutality, writings on the subject are shockingly thin on the ground. I found that it's usually mentioned only in passing—a page here, or a paragraph there. Apart from a few dozen articles by social psychologists, there is scarcely any literature on it at all.[4] If dehumanization really has the significance that scholars claim, then untangling its dynamics ought to be among our most pressing priorities, and its neglect is as perplexing as it is grave. I wrote this book to bring dehumanization out of the shadows, and to jump-start a conversation that is centuries overdue. To do this, I've drawn from a rich palette of sources—including history, psychology, philosophy, biology, and anthropology—to paint a portrait of dehumanization and the forces and mechanisms that sustain it.

It's sometimes said that dehumanization is a social construction that's at most a few centuries old. According to this story, dehumanization was, paradoxically, a child of the doctrine of universal human rights. This idea was the moral and political touchstone of the Enlightenment, but it conflicted with the brutal colonialism perpetrated by Europeans. As the example with which I began this chapter suggests, the dissonance between theory and practice was resolved by denying the humanity of the oppressed.

This story expresses a truth, but it is a partial truth that obfuscates the real nature, history, and extent of the dehumanizing impulse. Dehumanization is neither uniquely European nor uniquely modern. It is far more widespread, vastly more ancient, and more profoundly intertwined with the human experience than the constructionist view allows. To understand its workings, it's not good enough to examine some contingent facts about a particular historical period. We must look much deeper.

Of course, particular manifestations of dehumanization are social constructions, in the sense that they appear in a given culture and historical epoch which leave their distinctive stamp on it. Eighteenth-century Europeans embraced a certain type of dehumanization, but so did the Athenians during the fourth century before Christ, the Germans of the 1930s and '40s, and the Eipo tribesmen of highland New Guinea, who refer to their enemies as dung flies, lizards, and worms.[5]

In this book, I will argue that dehumanization is a joint creation of biology, culture, and the architecture of the human mind. Grasping its nature and dynamics requires that we attend to all three elements. Excluding any of them leaves us with a hopelessly distorted picture of what we are trying to comprehend.

Dehumanization is too important a topic to be left to the experts, so I've tried to make this book appealing and accessible to a broad general readership while at the same time addressing the concerns of specialist scholars in several disciplines. In doing this, I've done my best to balance academic rigor with engaging prose, on the principle that anything worth explaining is worth explaining in a clear and interesting way. In general, I've avoided technical jargon as much as possible, and have included explanations on the occasions when its use was unavoidable. However, there are a couple of exceptions to this rule. There are two ordinary words that I sometimes use in out-of-the-ordinary ways. The words are *person* and *human*. This isn't a self-indulgent plunge into academic obscurity. It's motivated by the need for a vocabulary to capture ideas that are hard to put into ordinary speech.

Let me explain . . .

Think of the word *dehumanization*. It literally means something like "removing the human-ness." Now, take someone and imagine that their humanity has been stripped away from them. What's left? When the founding fathers dehumanized their slaves, what remained of them? When European colonists dehumanized Native Americans or Nazis dehumanized Jews, what remained? In their eyes, what was left was a creature that seemed human—had a human-looking form, walked on two legs, spoke human language, and acted in more-or-less human ways—but which was nonetheless not human. As I will explain in detail later on, dehumanization is the belief that some beings only *appear* human, but beneath the surface, where it really counts, they aren't human at all. The Nazis labeled Jews as *Untermenschen* ("subhumans") because they were convinced that, although Jews looked every bit as human as the average Aryan, this was a facade and that, concealed behind it, Jews were really filthy, parasitic vermin. Of course, Jews did not wear their subhumanity on their sleeves. They were regarded as insidiously subhuman. Their ostensible humanity was, at best, only skin deep.

It's clear from these considerations we need a vocabulary to express the conceptual distinction between *appearing* human and *being* human. In defiance of the norms of common speech as well as time-honored academic convention, I reserve the word *person* for any being that appears human. You, the reader, are a person in this sense, and if Dracula, the Terminator, or any other man-shaped monster existed, they would be persons, too. I use *human* for beings that are members of our own kind, irrespective of their appearance (although what's meant by "our own kind" won't become fully clear until Chapter Seven). You are human, but Dracula and the Terminator aren't, even though they look human. John Merrick, the "Elephant Man" was human, in spite of his nonhuman appearance.*

I also want to address what may seem like a major omission. I have little to say about the role of dehumanization in the oppression of women. This is because the particular form of dehumanization that typically has been directed against women is fundamentally different from the form of dehumanization that I explore in this book. Since the 1980s, a number of feminist scholars, including Andrea Dworkin, Catherine MacKinnon, and Linda LeMoncheck, have argued that women are dehumanized by being objectified. When men objectify women they perceive them as *things* rather than human beings, as desirable lumps of flesh rather than human subjects.[6] In this book I am concerned with the kind of dehumanization associated with war, genocide, and other forms of mass violence. The objectification of women is produced by a different concatenation of forces, and its analysis demands a

---

*I use these words in this special way only in contexts where the distinction matters. Elsewhere, I stick to their vernacular meanings.

somewhat different set of conceptual tools. Apart from some dubious speculations by scholars working within a psychoanalytic framework, the psychological dynamics of objectification have been given short shrift in favor of its sociopolitical dimension. From time to time I gesture toward ways that a psychological analysis of dehumanization of women might be approached, but this is a large topic and requires a book of its own to do it justice.

Another omission concerns certain groups—sexual minorities (notably gay people), immigrants, mentally and physically handicapped people, and various specific ethnic groups (for example, the Roma, the Italians, and the Irish)—all of whom have, at one time or another, been victims of dehumanization. I say little or nothing about them in this book, not because I minimize their importance, but rather because the sheer pervasiveness of dehumanization made it impossible to discuss all of its manifestations while keeping the book to a readable length. So, I've had to deliberately restrict my focus. I've chosen to concentrate largely (but not exclusively) on the dehumanization of Jews, sub-Saharan Africans, and Native Americans, for a couple of reasons. One is their immense historical significance. The human story is filled with pain and tragedy, but among the horrors that we have perpetrated on one another, the persecution and attempted extermination of the Jewish people, the brutal enslavement of Africans, and the destruction of Native American civilizations in many respects are unparalleled. The other reason is that they have been richly documented, which makes them excellent paradigm cases for discerning the core features of the dehumanizing process. What we learn from them can then be applied elsewhere.

Having set the stage, and cleared up a few potential sources of misunderstanding, here is a preview of how the story will unfold.

In Chapter One, I explain why investigating dehumanization is worthwhile. To do this, I make use of some examples from World War II. Although most educated people are aware that the Nazis dehumanized Jews, Gypsies, and others, it's less commonly known that *all* the major players in the war, including the Allied forces, dehumanized their enemies. After delving into these historical examples, I talk about the role of dehumanization in the contemporary world, focusing on how it manifests in the mass media, particularly in coverage of the ongoing conflicts in the Middle East and the battle against terrorist organizations.

Most discussions of the history of the concept of dehumanization begin with the work of twentieth-century social psychologists. But these were latecomers: the story actually begins many centuries earlier. My mission in Chapter Two is to describe how the concept of dehumanization evolved over the centuries, starting with ancient authors such as Aristotle, Augustine, and Boethius, then moving forward through the Middle Ages and Enlightenment right up to the present. This hitherto unwritten piece of history also gives me an opportunity to introduce a key theoretical idea that will play an important role later on in the book: the notion of essence.

In Chapter Three, I tell the story of the colonization of the New World, and the dehumanization of its indigenous peoples. The question of whether Native Americans were human had been simmering ever since the Spanish arrived in the Caribbean. It came to a head in 1550, when campaigner for Indian rights Bartolomé de Las Casas clashed with Spanish humanist Juan Ginés de Sepúlveda in a debate that has been described as one of the most extraordinary events in Western political history.[7] I use these events as a springboard to discuss and assess ideas about dehumanization that have been advanced by psychologists since the early 1970s, and finally round off the chapter with a brief discussion of the notions of essence and appearance introduced in Chapter Two.

Chapter Four focuses on the role of dehumanization in slavery. I discuss the history of slavery, from ancient times onward, including both the trans-Saharan and the transatlantic slave trades, all the while focusing on how slaves were considered subhuman animals. I also touch on race and racism in this chapter (a subject to which I return in Chapter Six), and then turn to the issue of moral disengagement, looking at how dehumanization weakens our inhibitions against behaving cruelly toward our fellow human beings.

Chapter Five takes up the role of dehumanization in genocide. I survey the part it played in six major genocides: the German genocide of the Herero in 1904, the Armenian genocide of 1915–16, the Holocaust, the Cambodian genocide, the Rwandan genocide of 1994, and the recent genocide in the Darfur region of Sudan. I then go on to examine a Nazi publication of the 1940s entitled *The Subhuman*, and use this text to identify some of the core features of the dehumanizing process.

The next three chapters pull together many of the strands from the preceding chapters, and weave them into a theory of dehumanization that is sensitive to its cultural, psychological, and biological dimensions.

Chapter Six looks at the concept of race, and the connection between racism and dehumanization. Although everyday notions of race are scientifically groundless, most people continue to take the idea of race seriously. Social constructivists see race as an ideological category, but they ignore its psychological underpinnings. I argue that, understood correctly, the notion of race (together with the psychological processes responsible for our tendency to view people through racially tinted spectacles) is crucial for making sense of the dehumanizing process. Dehumanization feeds on racism; without racism, it probably couldn't exist.

It's often said that war is not uniquely human, and that ants as well as chimpanzees also wage war on one another. In Chapter Seven, I critically assess this assertion, and argue that same-species killing by these animals should *not* be considered a form of war, basing my conclusion on a comparison of intercommunity violence or "raiding" by chimpanzees and raiding by the Yanomamö of Amazonia. I then go on to explore Jane Goodall's claim that humans are the only animal capable of being cruel. I closely scrutinize the concept of cruelty and argue that Goodall is right. This leads to a deeper understanding of how dehumanization causes moral disengagement.

Chapter Eight has three parts. In the first, I focus on human beings' ambivalence about killing. On one hand we slaughter members of our own kind with gusto, while on the other hand we express a horror of spilling human blood. I argue that this isn't mere hypocrisy, but expresses two sides of human nature, each of which is authentic. In the second part, I examine how the capacity for dehumanization might have arisen in our species. I suggest that it was probably a by-product of several other developments, and that it was (and is) a solution to conflicts generated by the uniquely human ability to reflect upon our own thoughts.

Finally, in Chapter Nine, I recapitulate the major elements of the theory of dehumanization proposed in this book, and reflect on the question of where we need to look for solutions to the problem of dehumanization.

1
LESS THAN HUMAN

Palestine is our country.
The Jews our dogs.

—PALESTINIAN NURSERY RHYME

Arabs are the same as animals. There is no animal worse
than them.

—RABBI OVADIA YOSEF, *HAARETZ* [1]

"COME ON DOGS. Where are all the dogs of Khan Younis? Son of a
bitch! Son of a whore! Your mother's cunt!" Degrading taunts in Arabic
rang out from behind the fence that divided the Palestinian side of
the Khan Younis refugee camp from the Israeli side. Located near
the southern tip of the Gaza Strip, just outside the ancient town of
Khan Younis, the camp was established to house 35,000 of the nearly
one million Arabs who had been displaced by the 1948 Arab-Israeli
war. By the beginning of the twenty-first century its population had
swelled to over 60,000 souls housed in thirteen squalid cement
blocks.

The torrent of invective did not come from the mouth of an angry
Muslim; it was broadcast from a loudspeaker mounted on an armored
Israeli Jeep. *New York Times* journalist Chris Hedges was in the camp
that day, and watched as Palestinian boys began to lob stones at the Jeep
in a futile gesture of defiance. Hedges recounts:

There was the boom of a percussion grenade. The boys, most no more than ten or eleven years old, scattered, running clumsily through the heavy sand. They descended out of sight behind the dune in front of me. There were no sounds of gunfire. The soldiers shot with silencers. The bullets from M-16 rifles, unseen by me, tumbled end-over-end through their slight bodies. I would see the destruction, the way their stomachs were ripped out, the gaping holes in their limbs and torsos, later in the hospital.[2]

Four children were shot. Only three survived. One of them, a boy named Ahmed, explained to Hedges what had happened. "Over the loudspeakers the soldiers told us to come to the fence to get chocolate and money," he said. "Then they cursed us. Then they fired a grenade. We started to run. They shot Ali in the back."[3]

Khan Younis had long been a stronghold of Hamas, the Islamic Resistance Movement, and when the Israeli troops pulled out of the Gaza Strip in the fall of 2005, the bright green banners of Hamas fluttered from the asbestos rooftops of the camp. Hamas was founded in 1987 to end the Israeli presence in the region and to establish an Islamic state with Jerusalem as its capital. Although Hamas is devoted mainly to supporting schools, hospitals, and cultural activities, it is best known for its violence—its abductions, assassinations, suicide bombings, and rocket attacks against Israeli civilians. Osama Alfarra, the mayor of Khan Younis and a member of Hamas, was one of the many Palestinians who rejoiced when Israel relinquished control of the Gaza strip. "Gaza was a beginning," he told a reporter from the British *Guardian* newspaper. "You know how you hunt foxes? You dig them out of their holes. The fox is gone from Gaza to the West Bank. The resistance will dig him out of his hole there."[4]

Osama Alfarra and the anonymous soldiers in the Jeep stood on opposite sides of a single conflict. And yet, their attitudes were uncannily alike. Each portrayed the other as a nonhuman animal. The soldier represented Ali and his companions as dogs, unclean animals in both Jewish and Islamic lore. Likewise, Osama Alfarra's depiction of Israel as a fox represents a whole nation as vermin, fit to be hunted down and destroyed. The sly fox, an amalgam of greed and guile, has much in common with the traditional derogatory stereotype of the Jew, as exemplified by thirteenth-century Muslim writer Al-Jaubari's characterization of the Jewish people in *The Chosen One's Unmasking of Divine Mysteries*:

> Know that these people are the most cunning creatures, the vilest, most unbelieving and hypocritical. While ostensibly the most humble and miserable, they are in fact the most vicious of men. This is the very essence of rascality and accursedness. . . . Look at this cunning and craft and vileness; how they take other people's moneys, ruin their lives. . . .

And more recently, the remarks of Imam Yousif al-Zahar, a member of Hamas, conveyed the same idea. "Jews are a people who cannot be trusted," he remarked. "They have been traitors to all agreements—go back to history. Their fate is their vanishing."[5]

The soldier in the Israeli military Jeep dehumanized his Palestinian targets, and Osama Alfarra and his comrades dehumanized their Israeli enemies. In both examples—and in many, many more that I will describe in this book—a whole group of people is represented as less than human, as a prelude and accompaniment to extreme violence. It's tempting to see reference to the subaltern other as mere talk, as nothing more than degrading metaphor. I will argue that this view is sorely misguided. Dehumanization isn't a way of talking. It's a way of thinking—a way of thinking that, sadly, comes all too easily to us. Dehumanization is a scourge, and has been so for millennia. It acts as a psychological lubricant, dissolving our inhibitions and inflaming our destructive passions. As such, it empowers us to perform acts that would, under other circumstances, be unthinkable. In the pages and chapters to follow, I will do my best to explain what this form of thinking consists in, how it works, and why we so readily slip into it.

Before I get to work explaining how dehumanization works, I want to make a preliminary case for its importance. So, to get the ball rolling, I'll briefly discuss the role that dehumanization played in what is rightfully considered the single most destructive event in human history: the Second World War. More than 70 million people died in the war, most of them civilians. Millions died in combat. Many were burned alive by incendiary bombs and, in the end, nuclear weapons. Millions more were victims of systematic genocide. Dehumanization made much of this carnage possible.

Let's begin at the end. The 1946 Nuremberg doctors' trial was the first of twelve military tribunals held in Germany after the defeat of Germany and Japan. Twenty doctors and three administrators—twenty-two men and a single woman—stood accused of war crimes and crimes against humanity. They had participated in Hitler's euthanasia program, in which around 200,000 mentally and physically handicapped people deemed unfit to live were gassed to death, and they performed fiendish medical experiments on thousands of Jewish, Russian, Roma, and Polish prisoners.

Principal prosecutor Telford Taylor began his opening statement with these somber words:

> The defendants in this case are charged with murders, tortures, and other atrocities committed in the name of medical science. The victims of these crimes are numbered in the hundreds of thousands. A handful only are still alive; a few of the survivors will appear in this courtroom. But most of these miserable victims were slaughtered outright or died in the course of the tortures to which they were subjected. . . . To their murderers, these wretched people were not individuals at all. They came in wholesale lots and were treated worse than animals.[6]

He went on to describe the experiments in detail. Some of these human guinea pigs were deprived of oxygen to simulate high-altitude parachute jumps. Others were frozen, infested with malaria, or exposed to mustard gas. Doctors made incisions in their flesh to simulate wounds, inserted pieces of broken glass or wood shavings into them, and then, tying off the blood vessels, introduced bacteria to induce gangrene.

Taylor described how men and women were made to drink seawater, were infected with typhus and other deadly diseases, were poisoned and burned with phosphorus, and how medical personnel conscientiously recorded their agonized screams and violent convulsions.

The descriptions in Taylor's narrative are so horrifying that it's easy to overlook what might seem like an insignificant rhetorical flourish: his comment that "these wretched people were . . . treated *worse than animals.*" But this comment raises a question of deep and fundamental importance. What is it that enables one group of human beings to treat another group as though they were subhuman creatures?

A rough answer isn't hard to come by. Thinking sets the agenda for action, and thinking of humans as less than human paves the way for atrocity. The Nazis were explicit about the status of their victims. They were *Untermenschen*—subhumans—and as such were excluded from the system of moral rights and obligations that bind humankind together. It's wrong to kill a person, but permissible to exterminate a rat. To the Nazis, all the Jews, Gypsies, and the others were rats: dangerous, disease-carrying rats.

Jews were the main victims of this genocidal project. From the beginning, Adolf Hitler and his followers were convinced that the Jewish people posed a deadly threat to all that was noble in humanity. In the apocalyptic Nazi vision, these putative enemies of civilization were represented as parasitic organisms—as leeches, lice, bacteria, or vectors of contagion. "Today," Hitler proclaimed in 1943, "international Jewry is the ferment of decomposition of peoples and states, just as it was in antiquity. It will remain that way as long as peoples do not find the strength to get rid of the virus." Both the death camps (the gas chambers of which were modeled on delousing chambers) and the *Einsatzgruppen* (paramilitary death squads that roamed across Eastern Europe following in the wake of the advancing German army) were responses to what the Nazis perceived to be a lethal pestilence.[7]

Sometimes the Nazis thought of their enemies as vicious, bloodthirsty predators rather than parasites. When partisans in occupied regions of the Soviet Union began to wage a guerilla war against German forces, Walter von Reichenau, the commander in chief of the German army, issued an order to inflict a "severe but just retribution upon the Jewish subhuman elements" (the Nazis considered all of their enemies as part of "international Jewry," and were convinced that Jews controlled the national governments of Russia, the United Kingdom, and the United States). Military historian Mary R. Habeck confirms that, "soldiers and officers thought of the Russians and Jews as 'animals' . . . that had to perish. Dehumanizing the enemy allowed German soldiers and officers to agree with the Nazis' new vision of warfare, and to fight without granting the Soviets any mercy or quarter."[8]

The Holocaust is the most thoroughly documented example of the ravages of dehumanization. Its hideousness strains the limits of imagination. And yet, focusing on it can be strangely comforting. It's all too easy to imagine that the Third Reich was a bizarre aberration, a kind of mass insanity instigated by a small group of deranged ideologues who conspired to seize political power and bend a nation to their will. Alternatively, it's tempting to imagine that the Germans were (or are) a uniquely cruel and bloodthirsty people. But these diagnoses are dangerously wrong. What's most disturbing about the Nazi phenomenon is not that the Nazis were madmen or monsters. It's that they were ordinary human beings.

When we think of dehumanization during World War II our minds turn to the Holocaust, but it wasn't only the Germans who dehumanized their enemies. While the architects of the Final Solution were busy implementing their lethal program of racial hygiene, the Russian-Jewish poet and novelist Ilya Ehrenburg was churning out propaganda for distribution to Stalin's Red Army. These pamphlets seethed with dehumanizing rhetoric. They spoke of "the smell of Germany's animal breath," and described Germans as "two-legged animals who have mastered the technique of war"—"ersatz men" who ought to be annihilated.[9] "The Germans are not human beings," Ehrenburg wrote, ". . . If you kill one German, kill another—there is nothing more amusing for us than a heap of German corpses."

> Do not count days; do not count miles. Count only the
> number of Germans you have killed. Kill the German—this

is your old mother's prayer. Kill the German—this is what your children beseech you to do. Kill the German—this is the cry of your Russian earth. Do not waver. Do not let up. Kill."[10]

This wasn't idle talk. The *Wehrmacht* had taken the lives of 23 million Soviet citizens, roughly half of them civilians. When the tide of the war finally turned, a torrent of Russian forces poured into Germany from the east, and their inexorable advance became an orgy of rape and murder. "They were certainly egged on by Ehrenburg and other Soviet propagandists," writes journalist Giles McDonough:

> East Prussia was the first German region visited by the Red Army. . . . In the course of a single night the Red Army killed seventy-two women and one man. Most of the women had been raped, of whom the oldest was eighty-four. Some of the victims had been crucified. . . . A witness who made it to the west talked of a poor village girl who was raped by an entire tank squadron from eight in the evening to nine in the morning. One man was shot and fed to the pigs.[11]

Meanwhile, halfway across the world, a war was raging in Asia. Like their German allies, the Japanese believed that they were the highest form of human life, and considered their enemies inferior at best and subhuman at worst. American and British leaders were depicted with horns sprouting from their temples, and sporting tails, claws, or fangs. The Japanese labeled their enemies as demons (*oni*), devils (*kichiku*), evil spirits (*akki* and *akuma*), monsters (*kaibutsu*), and "hairy, twisted-nosed savages." Americans were *Mei-ri-ken*, a double entendre translated as "misguided dog."[12]

Japan pursued its military goals with extravagant and unapologetic savagery. Consider the systematic atrocities following the capture of the Chinese city of Nanjing in December 1937, where—for six weeks— soldiers killed, mutilated, raped, and tortured thousands of Chinese civilians. The details are set out Honda Katsuichi's book *The Nanjing*

Massacre: A Japanese Journalist Confronts Japan's National Shame.
Katsuichi reports that one former staff sergeant told him, "of soldiers disembowling pregnant women and stuffing hand grenades up women's vaginas and then detonating them." Another sergeant confessed that, because a crying infant disturbed him while he was raping the child's mother, he "took a living human child . . . an innocent baby that was just beginning to talk, and threw it into boiling water."[13] It's hard for the mind to encompass such horror. How can ordinary men (and they *were* ordinary men) do such things? Yoshio Tshuchiya, another Japanese veteran, tells us the answer. "We called the Chinese 'chancorro' . . . that meant below human, like bugs or animals. . . . The Chinese didn't belong to the human race. That was the way we looked at it."[14] Tshuchiya describes how he was ordered to bayonet unarmed Chinese civilians, and what it was that enabled him to comply with this order. "If I'd thought of them as human beings I couldn't have done it," he observed. "But . . . I thought of them as animals or below human beings." Shiro Azuma, who participated in the atrocities at Nanjing, told an interviewer that when the women were raped they were thought of as human, but when they were killed they were nothing but pigs.[15]

What about the Americans and their English-speaking allies? We were the good guys, weren't we? Allied personnel also dehumanized their enemies (as one soldier wrote in a letter home, "It is very wrong to kill people, but a damn Nazi is not human, he is more like a dog") but on the whole dehumanized the Germans less than they did the Japanese. Germans, after all, were fellow Anglo-Saxons—strapping blue-eyed boys who might just as well have grown up on farms in Oklahoma. But the Japanese were another story. A poll of U.S. servicemen indicated that 44 percent would like to kill a Japanese soldier while only 6 percent felt the same way about Germans.[16]

The "Japs" were considered animals, and were often portrayed as monkeys, apes, or rodents, and sometimes as insects (in Herman Wouk's novel *The Caine Muntiny*, they are described as "large armed ants"). In a typical example of xenophobic zeal, Australian general Sir Thomas Blamey told his troops in the Pacific theater, "Your enemy is a curious race—a cross between a human being and an ape. . . . You know that we have to exterminate these vermin if we and our families are to live. . . ." Pulitzer Prize–winning war correspondent Ernie Pyle, whose folksy dispatches were published in hundreds of newspapers, confirmed that this attitude was not limited to commanding offers, but was common among ordinary grunts as well. Pyle reported that "the Japanese were looked upon as something subhuman and repulsive, like cockroaches or mice." Ordinary citizens far away from the sharp edge of battle participated, too. The theme of the most popular float in an all-day parade in New York City in 1942 was bombs falling on a pack of yellow rats. It was named "Tokyo, We Are Coming."[17]

Viewing the Japanese as subhuman may have contributed to the practice of mutilating their corpses and taking their body parts from them as trophies. Charles Lindbergh recorded in his wartime diary that U.S. servicemen carved penholders and paper knives out of the thigh bones of fallen Japanese soldiers, dug up their decaying corpses to extract gold teeth, and collected ears, noses, teeth, and even skulls as wartime mementos. Taking human body parts as trophies was rare in the European theater. As military historian John Dower points out, if Allied troops had similarly mutilated German or Italian corpses this would have provoked an uproar.[18]

The dead weren't the only targets. Surrendering soldiers and prisoners were frequently killed and sometimes tortured. The philosopher and World War II veteran J. Glenn Gray recounts a revealing anecdote in this connection.

> An intelligent veteran of the war in the Pacific told a class of mine . . . how his unit had unexpectedly "flushed" a Japanese soldier from his hiding place well behind the combat area. . . . The unit, made up of relatively green troops, was resting and joking, expecting to be sent forward to combat areas. The appearance of this single enemy soldier did not frighten them. . . . But they seized their rifles and began using him as a life target while he dashed frantically around the clearing in search of safety. The soldiers found his

movements uproariously funny and were prevented by their laughter from making an early end of the unfortunate man. Finally, however, they succeeded in killing him, and the incident cheered the whole platoon, giving them something to joke and talk about for days afterwards.

To this Gray adds:

In relating this story to the class, the veteran emphasized the similarity of the enemy soldier to an animal. None of the American soldiers apparently even considered that he may have had human feelings of fear and wished to be spared.[19]

One of the most unsettling examples of dehumanization of the Japanese by Americans appeared in the U.S. Marine Corps' *Leatherneck* magazine. It's a brief piece, apparently intended to be humorous. Emblazoned across the top of the page is an illustration of a repulsive animal with a caterpillar-like body and a grotesque, stereotypically Japanese face, labeled *Louseus japanicus*. The text below it explains that the "giant task" of exterminating these creatures will only be complete when "the origin of the plague, the breeding grounds around the Tokyo area" are completely annihilated. The article was published in March 1945, the same month that U.S. aircraft rained incendiary bombs on Tokyo, burning up to 100,000 civilians alive. Over the next five months, around half a million noncombatants—men, women, and children—were, in the words of Major General Curtis LeMay, "scorched and boiled and baked to death" as sixty-seven Japanese cities were incinerated by fire bombs. And then, in August, nuclear weapons flattened Hiroshima and Nagasaki, with massive civilian casualties.[20]

DEHUMANIZATION IN THE MEDIA

I saw two naked detainees, one masturbating to another kneeling with its mouth open. I thought I should just get

out of there. I didn't think it was right . . . I saw SSG
Frederick walking toward me, and he said, "Look what
these animals do when you leave them alone for two
seconds."

—U.S. ARMY SPECIALIST MATTHEW WILSON,
TESTIMONY REGARDING ABU GHRAIB[21]

Dehumanization is aroused, exacerbated, and exploited by propaganda.
It's common knowledge that, prior to and during the 1994 genocide,
government radio broadcasts characterized Rwandan Tutsis as cockroaches, and that Nazi Germany had a propaganda apparatus devoted
to painting horrifying pictures of Jews and other supposed enemies of
the *Volk*. Russian political art from the 1930s and '40s portrayed German and Italian fascists and their allies as a veritable menagerie, including rats, snakes, pigs, dogs, and apes. And when fascists were depicted
in human form, they were endowed with subhuman attributes, like
pointed ears, fangs, or a nonhuman complexion. But apart from notorious examples like these, there is little awareness of the extent to which
the mass media are instrumental for propagating dehumanizing stereotypes.[22]

Journalists have always had an important role to play in disseminat-
ing falsehoods to mold public opinion, and this often involves dehumanizing military and political opponents. In a speech delivered at
London's Royal Albert Hall in 1936, against the background of the
gathering storm of fascism in Europe, Aldous Huxley argued that dehumanization is the *primary* function of propaganda.

> Most people would hesitate to torture or kill a human being
> like themselves. But when that human being is spoken of as
> though he were not a human being, but as the representative
> of some wicked principle, we lose our scruples. . . . All po-
> litical and nationalist propaganda aims at only one thing; to
> persuade one set of people that another set of people are not
> really human and that it is therefore legitimate to rob, swin-
> dle, bully, and even murder them.[23]

Collections of twentieth-century political posters confirm that visual propaganda from the United States, Germany, Britain, France, the Soviet Union, Korea, and elsewhere have often portrayed "the enemy" as a menacing nonhuman creature.[24] But you don't need to sift through historical archives to find examples of dehumanization in the popular media. All that you need to do is open a newspaper or turn on the radio.

On September 4, 2007, the *Columbus Dispatch* published a cartoon portraying Iran as a sewer with a swarm of cockroaches pouring out of it. The subtext wasn't subtle, and readers quickly got the message. "I find it extremely troubling that your paper would behave like Rwandan Hutu papers that also published cartoons depicting human beings . . . as cockroaches," one reader wrote, "calling for them to be stamped out—leading to genocide." Another wrote, "Depicting Iranians as cockroaches spewing out of a sewer was a vile slur on the Iranian people. . . . Cartoons like this only cause the neoconservative drums of war sounding for a disastrous military attack against Iran to beat louder."[25]

Three years earlier, when the Abu Ghraib prison scandal in Iraq became public, Rush Limbaugh—the most popular radio broadcaster in the United States, whose syndicated radio show has, at last count, 13 million listeners—described the prisoners who had been killed, raped, tortured, and humiliated by or at the behest of U.S. military personnel, as less than human. "They are the ones who are sick," fumed Limbaugh.

> They are the ones who are perverted. They are the ones who are dangerous. They are the ones who are subhuman. They are the ones who are human debris, not the United States of America and not our soldiers and not our prison guards.[26]

Limbaugh's view of the detainees was shared by members of the U.S. military establishment, including, presumably, their persecutors. The commander of Abu Ghraib, Brigadier General Janis Karpinski, subsequently disclosed that Major General Geoffrey Miller had told her to make sure that the prisoners were treated like animals. "He said they are like dogs and if you allow them to believe at any point that they are more than a dog then you have lost control of them," she said. (Miller had earlier "reformed" military interrogation techniques in Iraq along the lines used at Guantanamo Bay, and became deputy commanding general for detainee operations in Iraq after Karpinski's removal.)[27]

Michael Savage (the pseudonym of Michael Alan Weiner) is another popular radio host whose syndicated radio program is followed by 8 to 10 million listeners. Like Limbaugh, Savage derided the detainees as "subhuman" and "vermin," and suggested that forcible conversion to Christianity is "probably the only thing that can turn them into human beings." And in words uncannily reminiscent of Adolf Hitler's diatribes against the Jews, radio broadcaster Neal Boortz characterized Islam as a "deadly virus spreading through Europe and the West," adding, "We're going to wait far too long to develop a vaccine to find a way to fight this."[28]

Limbaugh, Boortz, and Savage play to the xenophobic gallery, so it's not surprising that they indulge in dehumanizing rhetoric from time to time. But this sort of talk is not confined to right-wing populists; it is well represented in mainstream media by journalists of all political stripes. Pulitzer prize–winning *New York Times* columnist Maureen Dowd wrote in a 2003 editorial that Muslim terrorists are "replicating and coming at us like cockroaches."[29]

Dehumanization makes strange bedfellows.

Newspaper headlines are a prime source of dehumanizing rhetoric. They're designed to catch the eye and to motivate you to read further. Describing human beings as bloodthirsty animals or dangerous parasites gets us to look because it plays on some of our deepest fears. Techniques like these arouse terror and close minds. If international conflicts are explained by the fact that our enemies are evil subhuman creatures, then no further analysis is needed.

Propaganda researchers Erin Steuter and Deborah Wills point out:

> The symbolic lexicon used by news media since 9-11 demonstrates a clear pattern. Suspected terrorists, enemy military and political leaders, and ultimately entire populations are metaphorically linked to animals, particularly to prey.

This holds true both nationally and internationally: headlines of newspapers of many political affiliations across the
US, Europe, and Australia generate, with remarkable con-
sistency, this journalistic framing of the enemy as hunted
animal. . . . [30]

Sometimes the wars in Iraq and Afghanistan are presented as a hunting expeditions ("As British close in on Basra, Iraqis scurry away"; "Terror hunt snares twenty-five"; and "Net closes around Bin Laden") with
enemy bases as animal nests ("Pakistanis give up on lair of Osama";
"Terror nest in Fallujah is attacked") from which the prey must be
driven out ("Why Bin Laden is so difficult to smoke out"; "America's
new dilemma: how to smoke Bin Laden out from caves"). We need to
trap the animal ("Trap may net Taliban chief"; "FBI terror sting nets
mosque leaders") and lock it in a cage ("Even locked in a cage, Saddam
poses serious danger"). Sometimes the enemy is a ravening predator
("Chained beast—shackled Saddam dragged to court"), or a monster
("The terrorism monster"; "Of monsters and Muslims"), while at other
times he is a pesky rodent ("Americans cleared out rat's nest in Afghanistan"; "Hussein's rat hole"), a venomous snake ("The viper awaits";
"Former Arab power is 'poisonous snake'"), an insect ("Iraqi forces find
'hornet's nest' in Fallujah"; "Operation desert pest"; "Terrorists, like rats
and cockroaches, skulk in the dark"), or even a disease organism ("Al
Qaeda mutating like a virus"; "Only Muslim leaders can remove spreading cancer of Islamic terrorism"). In any case, they reproduce at an
alarming rate ("Iraq breeding suicide killers"; "Continent a breeding
ground for radical Islam").[31]
Do you think that I'm making too much of these examples? Perhaps
they're only metaphors—just colorful ways of speaking and writing that
shouldn't be taken to imply that anyone is regarded as subhuman. You
may have noticed that even Steuter and Wills explicitly describe them
as metaphors. True, sometimes this sort of language *is* metaphorical—
but it's foolish to think of it as *just* metaphorical. Describing human
beings as rats or cockroaches is a symptom of something more powerful
and more dangerous—something that's vitally important for us to un-

derstand. It reflects how one *thinks* about them, and thinking of a person as subhuman isn't the same as calling them names. Calling people names is an effort to hurt or humiliate them. It's the use of language as a weapon. But dehumanizing a person involves judging them to be less than human. It's intended as a description rather than as an attack, and as such is a departure from reality—a form of self-deception. Whatever one's opinion of one's nation's enemies, the fact remains that they are human beings, not subhuman animals.

So far, most of my examples have been plucked from recent history. But dehumanization is far more widespread than that. It is found across a far-flung spectrum of cultures and appears to have persisted through the full span of human history, and perhaps into prehistory as well. It appears in the East and in the West, among sophisticates of the developed world and among remote Amazonian tribes. Its traces are inscribed on ancient cuneiform tablets and scream across the headlines of today's newspaper. Dehumanization is not the exclusive preserve of Nazis, communists, terrorists, Jews, Palestinians, or any other monster of the moment. We are all potential dehumanizers, just as we are all potential objects of dehumanization. The problem of dehumanization is *everyone's* problem. My task is to explain why.

2

STEPS TOWARD A THEORY OF DEHUMANIZATION

It wasn't only wickedness and scheming that made people unhappy, it was confusion and misunderstanding; above all, it was the failure to grasp that other people are as real as you.

—IAN MCEWAN, *ATONEMENT: A NOVEL*

THE WORD *DEHUMANIZATION* ENTERED the English lexicon early in the nineteenth century. From the outset, it had many meanings, and it still does today. Articles in both the popular press and in scholarly journals tell us that automatic ticketing machines in airports dehumanize customers by "turning them into cattle," that pornography dehumanizes women, that triathlons dehumanize athletes, that technology dehumanizes education, and that the treatment of prisoners dehumanizes them. This is just a small selection of the wide variety of ways that the word *dehumanization* is used.[1]

This unruly tangle of meanings poses a problem for anyone wanting to study dehumanization. To talk meaningfully about dehumanization, we need to pin it down. In this book, I use the term to refer to the act of conceiving of people as subhuman creatures rather than as human beings. This definition has two components: When we dehumanize people we don't just think of them in terms of what they lack, we also think of them as creatures that are less than human.

To make this clear, it's useful to contrast my concept of dehumanization with its most common alternatives.

It's sometimes said that people are dehumanized when they're not recognized as individuals. This happens when they are treated as numbers, mere statistics, cogs in a bureaucratic machine, or exemplars of racial, national, or ethnic stereotypes, rather than as unique individuals. This isn't what I mean by dehumanization. Taking away a person's individuality isn't the same as obliterating their humanity. An anonymous human is still human.

In other contexts, dehumanization is equated with *objectification*. This is the topic of feminist philosopher Linda LeMoncheck's book *Dehumanizing Women*, revealingly subtitled *Treating Persons as Sex Objects*.[2] The feminist notion of objectification comes to us from the writings of the Enlightenment philosopher Immanuel Kant via late twentieth-century feminist theory as developed by Catherine MacKinnon and Andrea Dworkin. Here's how Dworkin defines it:

> Objectification occurs when a human being, through social means, is made less than human, turned into a thing or commodity, bought and sold. When objectification occurs, a person is depersonalized, so that no individuality or integrity is available socially or in what is an extremely circumscribed privacy. Objectification is an injury right at the heart of discrimination: those who can be used as if they are not fully human are no longer fully human in social terms; their humanity is hurt by being diminished.[3]

When women are objectified, their humanity is disregarded. They are treated as instruments of sexual pleasure rather than as human subjects. However, treating someone as only a means to a sexual end is not the same as regarding them as subhuman, for one can fail to acknowledge a person's subjectivity without denying the existence of that subjectivity, just as one might not believe that it is raining without believing that it is *not* raining. This isn't just a word game. A surgeon disregards the humanity of the person lying on the operating table. He is interested in the patient as a flesh-and-blood machine in need of repair, not as a human subject, but in doing so he doesn't think of his patient as *less* than human.

At times it's said that we dehumanize people by taking a derogatory attitude toward them. But denigrating others falls short of denying their humanity. Often, it involves judging them to be inferior human beings rather than subhuman animals. An inferior human is still human.

Finally, dehumanization is sometimes equated with cruel or degrading treatment. It's said, for instance, that torturing a person, or systematically disrespecting them, is tantamount to dehumanizing them. This puts the cart before the horse. Doing violence to people doesn't make them subhuman, but conceiving of people as subhuman often makes them objects of violence and victims of degradation. The important thing to keep in mind is that dehumanization is something psychological. It occurs in people's heads. It's an attitude—a way of thinking about people—whereas harming them is a form of behavior, a kind of *doing* rather than a kind of *thinking*.

To dehumanize a person is to regard them as subhuman. This is how Abraham Lincoln used the word in his final debate with Stephen Douglas. The Lincoln/Douglas debates revolved around the issue of slavery. Douglas asserted that the Founding Fathers did not have "inferior or degraded" races in mind when they spoke of the equality of men.[4] Lincoln responded that Douglas displayed "the tendency to dishumanize the man" (or, in some reports, "dishumanize the negro") and thereby "take away from him all right to be supposed or considered as human." When the *New York Tribune* published his speech, the editors changed his awkward "dishumanize" to the more elegant "dehumanize."[5]

Dehumanization, as I have defined it, raises a multitude of questions, three of which are especially important. First, it invites us to consider what it really means to think of someone as a human being. Clearly, being human can't be the same as *looking* human, for if that were the case then Stephen Douglas would have acknowledged that African Americans are fully human. This implies that humanity runs deeper than meets the eye. If being human isn't the same as looking human, there must be something more subtle and less tangible at issue. This invites us to ponder the question of what exactly it is that dehumanized people are supposed to lack. Finally, we need to address the question of precisely what sort of creatures dehumanized others are supposed to be.

Dehumanization also raises many questions about the specifics of our psychology. What is it about human nature that enables us to conceive of one another as less than human? How, exactly, does dehumanization work, and why does it occur? Does it have a function, and if so, what is that function? Is the dehumanizing impulse universal, or is it culturally and historically specific? Is it a hard-wired product of our biological evolution, or is it learned? What are the typical patterns of the dehumanizing imagination, and why just these particular patterns?

A good theory of dehumanization should address all of these questions. I will do so in the chapters to follow, but I'm going to have to build up to them gradually. In this chapter I'll make a start by telling the story of how the concept of dehumanization developed, from the time of the ancient Greeks to the twenty-first century. It's a fascinating story, and one that's never before been told.

CLASSICAL ANTIQUITY: MAINLY ARISTOTLE, AUGUSTINE, AND BOETHIUS

Now what's going to happen to us without barbarians?
Those people were a kind of solution.
— CONSTANTINE CAVAFY, "WAITING FOR
THE BARBARIANS"[6]

The phenomenon of dehumanization is very ancient. However, the earliest attempts to explain it that have come down to us are only around 2,600 years old, and are found in the writings of the ancient Greeks. When the Greeks described people as subhuman animals, they often meant this to be taken at face value. Israeli classicist Benjamin H. Isaac points out that we find "a rich and varied literary tradition that uses animals as a literary device" in the ancient world, but we should not be tempted to regard this as just a rhetorical conceit. Isaac cautions us that "not all literary passages that represent people as animals should be interpreted as comparisons or metaphors."

> Some of them were intended quite literally. . . . [T]he ani-
> mal comparison was part of an attitude of mind, a way of
> thinking about oneself as distinct from a foreigner, which
> formed the framework in which imperialism could flourish
> unfettered by moral inhibitions or restraints.[7]

A seventh-century poem by Semonides of Amorgos explicitly describes women as subhuman creatures. The poem presents a taxonomy
of ten kinds of women, each of which is made from a different kind of
animal. "In the beginning," Semonides wrote, "the god made the female mind separately."

> One he made from a long-bristled sow. In her house, every-
> thing lies in disorder, smeared with mud and rolls about the
> floor; and she herself unwashed, in clothes unlaundered, sits
> by the dungheap and grows fat. Another he made from a
> bitch, vicious, own daughter of her mother, who wants to
> hear everything and know everything. . . . [8]

In addition to sows and bitches, he also describes women made
from vixens, asses, ferrets, mares, and monkeys. Although this may sound
like wordplay to twenty-first-century ears, Semonides probably had
something else in mind. Isaac points out that Semonides "claims that
the specific type of woman is literally made out of an animal . . . the
woman made from a sow *is* a sow."[9] The poem thus supplies a primitive theory of dehumanization: some people are subhuman, and their
subhumanity comes from the animal substance from which they are
made, whereas the highest type of human beings were thought to
be *autochthonous*—to have sprung from the soil of their homeland.
A similar theory (this time not restricted to women) was advanced
about a century later by Aesop, the Greek slave who authored *Aesop's
Fables*.

> Following Zeus' orders, Prometheus fashioned humans and
> animals. When Zeus saw that the animals far outnum-

bered the humans, he ordered Prometheus to reduce the number of animals by turning them into people. Prometheus did what he was told, and as a result those people who were originally animals have a human body but the soul of an animal.[10]

This may be the earliest written reference to the idea that it is possible to look just like a human being but have a subhuman soul.

As intriguing as these fragments are, the theory of dehumanization only came into its stride with the work of Aristotle during the fourth century BCE. The Greeks of Aristotle's era divided people into two categories: themselves and everybody else. They considered themselves to be paragons of civilization, and labeled all foreigners as *barbaroi* (barbarians). Aristotle's remarks on dehumanization lent grass-roots xenophobia a veneer of intellectual sophistication. He claimed that barbarians were *slaves by nature.*

Aristotle based his thesis on the idea that it's rationality that makes us human. His argument was premised upon a particular conception of biological purpose. Aristotle recognized that there's something that parts of organisms are for: they have a purpose. We all recognize that eyes are for seeing, hearts are for pumping blood, protective coloring is for evading predators, and wings are for flying. Notice that two things can have the same purpose even if they are quite dissimilar in other respects. Bird wings, insect wings, and bat wings are all very different. Birds' wings and bats' wings are modified forelimbs, but birds' wings are relatively rigid whereas bats' wings are modified hands that "grasp" the air. Butterfly wings aren't forelimbs at all. Each kind of wing is anatomically distinct from the others, and each has a different evolutionary history (of course, Aristotle didn't know anything about evolution; the example is mine, not his). So how come they're all wings?

The Aristotelian answer is that they are all wings because they share a common purpose: they're all for flying; oversimplifying somewhat, being for flying is what *makes* something a wing.[11]

Aristotle applied the same pattern of reasoning to human beings, arguing that there must be a purpose to being human—there's something that we are *for* that defines what we are, just like being for flying defines what it is to be a wing. He concluded that rationality is the defining characteristic of the human. Living thoughtfully and deliberately is the proper aim of human life. Human beings are for being rational like wings are for flying, and just as being for flying makes something a wing, so being rational makes one human.

I now need to introduce a theoretical notion that will be central to the theory of dehumanization developed later in this book: the notion of *essence*.

The philosophical concept of essence is derived from Aristotle, who used the odd expressions *"to ti ên einai"* ("the what it was to be") and *"to ti esti"* ("the what it is") at various points in his writings. His Roman translators were stumped for a Latin synonym, so they coined the term *essentia*, which they derived from the Latin verb *esse*, "to be." By the late fourteenth century, *essentia* had become anglicized as "essence."

As Aristotle's original terminology implies, the essence of a thing is that which makes it what it is. Essence contrasted with appearance (how things seem rather than what they are). Appearances are, so to speak, only skin deep, whereas essences cut to the core.

Philosophers often use the example of water to illustrate the distinction between essence and appearance. What exactly is water? You might be tempted to answer this question by reciting a description of water: "It's a clear, colorless, tasteless liquid that freezes at 0 degree centigrade, boils at 100 degree centigrade (at sea level), flows in rivers, comes out of showers," and so on. It's perfectly true that anything that is water has all of these characteristics. But do they pin down what water actually is?

Chemists don't define water in this way at all. In chemistry, water is defined by what it's made from. Water is H_2O.[12] Some stuff might have all of the superficial properties of water but still not be water. Vodka is a clear, colorless liquid—but it certainly isn't water! But what about the entire set of characteristics that we associate with water (all the characteristics mentioned in the preceding paragraph)? Don't they add up to a definition of what water is? Not exactly. As far as we know, everything that fits the description of water is H_2O, but there are circumstances in which the two could come apart. There could be a substance that's just like water in all these superficial respects, but which doesn't have the molecular structure of H_2O. This is at least conceivable. But it is inconceivable that scientists might discover a substance that's H_2O but isn't water.

We owe this insight to Harvard philosopher Hilary Putnam, who explained it by way of a famous thought experiment that's become known as "Twin Earth."[13] Putnam invited his readers to imagine that there's a planet just like Earth except for the fact that the stuff that flows in rivers and comes out of showers isn't H_2O, but XYZ (strange elements found in that galactic neighborhood). XYZ looks, tastes, smells, and behaves just like H_2O does. In fact, the residents of Twin Earth (who have never encountered H_2O) call it "water." So, Earthlings and Twin-Earthlings use the very same word for substances that are superficially indistinguishable but made out of completely different stuff. Now, suppose you took a vacation on Twin Earth and, while soaking in a hot tub, you entertained the thought that you were soaking in hot water. According to Putnam, this thought would be mistaken, because although the liquid in which you were pleasantly immersed was superficially indistinguishable from water, *it wasn't water*. It had the appearance of water (clear, colorless, tasteless, etc.) but lacked its essence. H_2O is the essence of water because it *makes* something water, whereas being a clear, colorless, tasteless liquid only makes something *resemble* water. If your Twin Earth doppelganger were luxuriating in a hot tub on Earth, and had the same thought, she would likewise be wrong, because in Twin Earthspeak, water is XYZ.

The distinction between essence and appearance isn't just an academic matter. It reflects the way we ordinarily think about things. It's what philosophers call an *intuition*. In the vernacular, when people talk about intuitions, they usually mean something spooky like telepathy and clairvoyance, but philosophers (and increasingly, psychologists) mean something completely different by the word. A philosophical intuition is just *how something seems* to you. Right now, you've got an intuition that you are reading this book. You don't need to *figure out* that you're reading it; it just seems obvious—directly disclosed to your consciousness—that you are reading it; and it is doubtful that anybody could convince you otherwise. Likewise, there's a growing body of psychological research (some of which I will explore in Chapter Six) showing that people intuit the dichotomy between essence and appearance. We spontaneously imagine that many things (for example, living things) have something "inside" them that makes them the kind of thing that they are, and that this isn't always reflected in how they look. Even small children assume that there is some inner property, which, for want of a better word, we can call "tigerness," that accounts for something's being a tiger, and are aware that a tiger with only three legs, or that's painted purple, or that has no tail—a tiger that lacks the stereotypical appearance of a tiger—is still a tiger.[14]

Thanks to our essentialistic proclivities, the idea that every human being is endowed with a human essence, an inner core (a soul, spirit, or distinctive genetic signature) is intuitively compelling. If you share this intuition, as most people seem to, then you will be open to the idea that someone can *be* human even though they don't *look* human (think of John Merrick, the "Elephant Man," whose physical deformities were so extreme that he looked like a creature belonging to a different species, even though he was a human being). On the flip side, you will probably find it credible that a creature could closely resemble a human being without being truly human. This is a popular theme in horror and science fiction, from Bram Stoker's *Dracula* to Arnold Schwarzenegger in *The Terminator*. Both Dracula and the Terminator have human forms and behave in more or less human ways (apart from certain idiosyncrasies), but neither of them is human on the "inside," where it counts.

Now, let's get back to Aristotle.

If, as Aristotle claimed, the ability to reason is what makes us human, any being that is unable to reason must be less than human. Aristotle believed that barbarians had only a rudimentary ability to reason, and this assumption underpinned his view that barbarians are natural slaves. Unlike the beasts of the field, the *barbaroi* could understand and respond to rational discourse, but they lacked the wherewithal to actively pursue it.

We may thus conclude that all men who differ from others as much as the body differs from the soul, or an animal from a man (and this is the case with all whose function is bodily service, and who produce their best when they supply such service)—all such are by nature slaves. In their case . . . it is better to be ruled by a master. Someone is . . . a slave by nature if . . . he participates in reason to the extent of apprehending it in another, though destitute of it in himself. Other animals do not apprehend reason but obey their instincts. Even so there is little divergence in the way they are used. Both of them (slaves and tame animals) provide bodily assistance in supplying essential needs. . . . It is thus clear that, just as some are by nature free, so others are by nature slaves, and for these latter the condition of slavery is both beneficial and just.[15]

Aristotle didn't quite assert that barbarians are subhuman, but he came uncomfortably close. He asserted that they are "incomplete" humans, rather than nonhuman animals. However, as is evident from the passage just quoted, he believed that barbarians had something in common with subhuman creatures, and also recommended war as a method for acquiring those natural slaves that stubbornly refuse to enter into the state of subjugation that is their proper destiny. In doing so, he chillingly compared warfare with hunting.

[W]arfare is by nature a form of acquisition—for the art of hunting is part of it—which is applied against wild animals and against those men who are not prepared to be ruled even though they are born for subjugation, in so far as this war is just by nature.[16]

In the Aristotelian scheme, barbarians are poised precariously on the cusp between humanity and subhumanity. Although a higher form of life than cattle, they are unable to reason. They are thus strangers to the civilized life of the *polis*, and can achieve human status only vicariously by submission to their fully human masters.

The theory of natural slavery had an immense impact on medieval thought, both in the Islamic world and in Christian Europe. The idea that certain people are naturally suited for slavery was to have tragic historical reverberations for many centuries after Aristotle's death in 322 BCE. It was invoked by Muslim thinkers like the philosopher Ibn Sina (aka Avicenna) and the jurist al-Andalusī to justify the horrors of the trans-Saharan slave trade, and by the Spanish *conquistadores* and English colonists to justify the conquest and enslavement of Native Americans (I will explore these topics in Chapter Four).[17]

The distinction between essence and appearance was an important component of philosophical thought, and in conceptions of subhumanity, long after Aristotle's death. Thus, the fifth-century philosopher and theologian Aurelius Augustinus Hipponensis, an Algerian Berber better known as Saint Augustine, assured his fellow Christians that a person's physical appearance has no bearing at all on their human status and spiritual worth.

> [W]homever is anywhere born a man, that is, a rational, mortal animal, no matter what unusual appearance he presents in color, movement, sound, nor how peculiar he is in some power, part, or quality of his nature, no Christian can doubt that he springs from that one protoplast. We can distinguish the common human nature from that which is peculiar, and therefore wonderful.

He continued in a vein that would do any twenty-first-century multiculturalist proud, asserting that "God . . . sees the similarities and diversities that can contribute to the beauty of the whole."[18]

Aristotle and Augustine both denied that looking fully human is the same as being human, but they approached it from opposite directions. Aristotle urged that a being that is indistinguishable from a fully fledged human being may yet be less than fully human, whereas Augustine proposed that peculiarities of appearance have no bearing on one's humanity. In spite of their differing emphases, these positions are two sides of a single coin. Both are derived from the distinction between appearance and essence.

The next important contribution to the theory of dehumanization came from Boethius, a Roman philosopher born into an aristocratic Christian family around the year AD 480, about fifty years after Augustine's death. While Boethius was still in his forties, and at the apex of a stellar career in government and scholarship, he was arrested, charged with treason, and sentenced to death. Sitting in a cell on death row, awaiting execution, Boethius wrote his masterpiece, *The Consolation of Philosophy*, which became one of the most widely read, and influential, works in European literature for the next millennium.

The *Consolation* unfolds as a conversation between Boethius and Lady Philosophy, a spiritual guide who leads him from ignorance to enlightenment, teaching him that misfortune is a blessing and that earthly happiness is ephemeral and illusory. At one point in their dialogue, Lady Philosophy explains to Boethius that wicked people lose their humanity. "All that falls away from the good," she assures him, "ceases also to exist, wherefore evil men cease to be what they were." She goes on to say that such men have "lost their human nature" and that "you cannot hold him to be a man who has been . . . transformed by his vices."

Even though evil people resemble human beings, they have lost the inner spark that makes one human. But if wicked people aren't really humans, then what sort of beings are they? Boethius approaches his answer obliquely, by way of a discussion of the role of animal imagery in everyday speech.

> If a violent man and a robber burns with greed of other men's possessions, you say he is like a wolf. Another fierce man is always working his restless tongue at lawsuits, and you will compare him to a hound. Does another delight to spring upon men from ambushes with hidden guile? He is as a fox. Does one man roar and not restrain his rage? He would be reckoned as having the heart of a lion. Does another flee and tremble in terror where there is no cause of fear? He would be held to be as a deer. If another is dull and lazy, does he not live the life of an ass? One whose aims are inconstant and ever changed at his whims, is in no wise different from

the birds. If another is in a slough of foul and filthy lusts, he
is kept down by the lusts of an unclean swine. Thus then a
man who loses his goodness, ceases to be a man, and . . .
turns into a beast.[19]

Wicked people repudiate their human essence, and acquire the essence of a nonhuman animal. To appreciate this, it helps to set aside
your twenty-first-century ways of thinking, and feel your way into
Boethius's mind-set. If a thing's essence defines what it is, then anything that loses its essence is no longer the thing that it was. In horror
films, when people become vampires, they forfeit their humanity;
they are transformed into a different sort of creature. Changes in a thing's
appearance have no such drastic effects. We are quite accustomed to
things changing their appearance. People might become pale, or dye
their hair, or lose ten pounds, but in each case, they remain the person that they were before. But suppose it were possible for them to
lose their humanity—to literally cease being human (for instance, by
becoming a vampire). If that could happen, it would be correct to say
that the person *ceased to exist*, and was seamlessly replaced by something else: someone (or some*thing*) with the very same appearance
but quite a different essence.

This is perhaps not quite as incredible as it might at first seem. Physics provides us with examples of things of one kind spontaneously mor-
phing into things of another kind. This weird process happens in
radioactive decay. Uranium (U^{238}) is a naturally occurring radioactive
metal, which becomes thorium (Th^{234})—an entirely different element,
possessing a different chemical essence—by spontaneously emitting alpha particles (particles composed of two neutrons and two protons).
When a uranium atom emits an alpha particle, *it ceases being uranium*.
In fact, the decay of uranium into thorium is only the first step of a long
"decay chain" consisting of thirteen more transformations before eventually finally becoming lead. Radioactive decay provides a nice analogy
for Boethius's theory of dehumanization. In becoming wicked, a person
loses his human essence. This causes him to cease being what he was:
a human being. The new entity is made from the same material as the

former one (minus the part that was lost), but is nevertheless a being of a completely different kind: a subhuman being with a new, subhuman, essence.

To grasp this idea more fully, it's necessary to wrap one's mind around a concept that Boethius and his contemporaries took for granted (and that most of us, often in spite of ourselves, tacitly embrace today): the concept of the great chain of being.

THE GREAT CHAIN OF BEING

Since ancient times, people have conceived of the universe as a vast hierarchy with God, the supremely perfect being, sitting astride its apex, with inanimate matter lying at its base and everything else situated at one or another of the many levels arrayed in between. Although the details of the scheme vary from one culture to another and from one epoch to the next, all versions of it are broadly similar. Plants are near the bottom, not much higher than the soil from which they grow. Simple animals like worms and snails are more perfect than plants, so they occupy a slightly more elevated rung. Mammals are higher still, and we humans have a privileged rank just below the angels, and two steps beneath the Creator. This taxonomic system was called the *great chain of being* (or *scala naturae*, literally "ladder of nature"). Alexander Pope succinctly described its attributes in his 1733 *Essay on Man*.

> *Vast chain of being! Which from God began,*
> *Natures aethereal, human, angel, man,*
> *Beast, bird, fish, insect, what no eye can see,*
> *No glass can reach; from Infinite to thee,*
> *From thee to nothing.—On superior pow'rs*
> *Were we to press, inferior might on ours;*
> *Or in the full creation leave a void,*
> *Where, one step broken, the great scale's destroy'd;*
> *From Nature's chain whatever link you strike,*
> *Tenth or ten thousandth, breaks the chain alike.*[20]

Philosopher Arthur O. Lovejoy, who wrote the classic and still definitive study of the great chain of being, believed that the idea originated in the ideas of Plato and Aristotle as synthesized by the Egyptian philosopher Plotinus during the third century. However, Lovejoy overestimated the intellectual influence of Greek philosophy for, although the version of the great chain that proliferated in both Western and Islamic societies was strongly influenced by Greek philosophical ideas, its pedigree is much older and its prevalence more widespread. In the *Book of Genesis*, for instance, we are told that God (*Elohim*) made human beings in his own image, and destined them to have dominion over all other living things (just as God has dominion over him) and we find a similar principle in Eastern ideas of the ascent of the soul from animal forms to the divine through cycles of birth, death, and reincarnation. In fact, the idea can be traced right back to the ancient Egyptian and Mesopotamian civilizations, before disappearing into the darkness of prehistory.[21]

In today's secular world we segregate fact from value, and our scientific accounts of the structure of the universe are purely descriptive. This point of view would have been alien to Boethius and others who endorsed the idea of the great chain of being, for the *scala naturae* seamlessly melded fact with value. The position of anything on the hierarchy was an index of its intrinsic worth. It was also believed that each level of the chain contains its own hierarchy of sublevels and sub-sublevels. For instance, in the category of metals (itself a subcategory of inanimate substances), gold was considered the most noble and lead the most base, so the former belonged at the top of the hierarchy of metals, while the latter occupied a humble position at the bottom. The hierarchical notion of sublevels sparked a great deal of speculation about the relative superiority or inferiority of different types of human beings. The twelfth-century philosopher Albertus Magnus (Saint Albert the Great) claimed that there is a category of *similitudiens hominis* (creatures similar to man, in which he included apes and pygmies) lying in between humans and animals. But this didn't stick. Later, scholars jettisoned Albert's innovation, and divided humanity into a series of subtypes ranked from "highest" to "lowest." Unsurprisingly, considering their origin, most of these schemes modestly placed Caucasians at the pinnacle of humanity and relegated Native Americans and sub-Saharan Africans to the bottom, only a hair's breadth away from the apes (an idea that underpinned Thomas Jefferson's infamous claim that male apes preferred African women as mates to members of their own species).[22]

The great chain of being represented the cosmos as static and unchangeable, complete and continuous. There was no room for novelty. How different this is from the Darwinian picture of constant flux, in which machinery of nature eternally generates "forms most beautiful and most wondrous" and consigns others to oblivion. Perhaps the most revolutionary aspect of Darwin's work lay in his replacing the ancient, vertical model of biodiversity with a more egalitarian, horizontal one. He denied that life progresses toward a goal of perfection and asserted instead that it simply diversifies—branching outward in a ramifying network of taxa rather than climbing upward. Darwin resisted adopting the word *evolution* to characterize his theory precisely because of its connotation of progress towards a predetermined goal. But even now, in today's post-Darwinian world, we find ourselves clinging to the more ancient vision of the cosmos. We still unblushingly speak of organism being "higher" or "lower" on an evolutionary scale, and the assumption that our species is more highly evolved than others continues, after all these centuries, to suffuse the zeitgeist. Even the scientific term *primate*, which refers to the order of animals that includes *Homo sapiens*, comes from the Latin *primas*, meaning "of the highest rank."

Why is the human imagination so thoroughly captured by the metaphor of the great chain of being? Perhaps we cling to it because (status-obsessed primates that we are) we project the socially stratified character of human societies, with their hierarchies of wealth, class, and power, onto the cosmos. Or perhaps it's inspired by the fear that moral values would evaporate if we denied that some beings have greater intrinsic worth than others. John Stuart Mill's remark, that it is better to be a dissatisfied human being than a satisfied pig, is interesting in this connection, because it tacitly relies on the notion of a natural hierarchy to underwrite a moral theory. Mill wanted to base his ethical theory on pleasure. His utilitarian doctrine had it that what's morally good is nothing more than whatever maximizes human pleasure. But Mill couldn't bring himself to accept that all pleasures are created equal, so he illicitly invoked the idea of "higher" and "lower" pleasures, which undermined the whole idea that moral value could be analyzed in terms of pleasure by sneaking moral rectitude in through the back door. Mill associated higher and lower pleasures with higher and lower forms of life, hence his remark about humans and pigs. But why should a pig's satisfaction be less worthy than a human's dissatisfaction? Like so many of us, Mill couldn't manage to liberate himself from the notion that pigs are beneath us in the cosmic order.

The great chain of being continues to cast a long shadow over our contemporary worldview. It's also a prerequisite for the notion of dehumanization, for the very notion of *sub*humanity—of being *less* than human—depends on it.

Boethius pictured the world in ways that seem very strange from today's perspective. Later thinkers would not countenance the idea that human beings literally become transformed into subhumans, no matter how depraved they are. Instead, they would look for psychological explanations of why we *imagine* that others are less than human. However, his emphasis on the idea that a person can lose his or her human essence captured something very important—indeed, essential—about the way that we think about dehumanization.

THE MIDDLE AGES: ISLAM, PICO, AND PARACELSUS

Reputation, reputation, reputation! Oh, I have lost my reputation. I have lost the immortal part of myself, and what remains is bestial.

—WILLIAM SHAKESPEARE, *OTHELLO*[23]

Surely the vilest of animals in Allah's sight are those who disbelieve.

—THE KORAN[24]

Boethius was a liminal figure, who can be seen as either one of the last great thinkers of classical antiquity, or one of the first of the Middle Ages. He lived during the twilight years of the Roman Empire, when the great cultural legacy of Greco-Roman civilization was taking its last few dying gasps. As the curtain fell, and Europe was plunged into the Dark Ages, a new religious movement was stirring in the Arabian Peninsula. Within a century of Muhammad's death in 632, Islamic civilization had burst out of its birthplace and created a vast empire that stretched from Spain and southern France in the west, all the way to central Asia in the east.

From its inception, dehumanization had a place in the Muslim conception of the world. Medieval Muslims took it for granted that humans can become subhuman. It was underwritten by the authority of the Koran, as well as several *hadith* (sayings attributed to Muhammad). Unlike Aristotle and Boethius, the early Muslims viewed reversion to the subhuman state as a form of divine punishment.

Early Muslim references to dehumanization were overtly ethnocentric. Almost without exception, the people who are transformed into subhuman creatures—specifically, pigs, apes, and rats—are Jews.

There are three verses in the Koran that describe the transformation of Jews into nonhuman animals as punishment. They are punished not because they are Jews, but rather because they transgressed the injunction in the Torah to refrain from working on the Sabbath.

> And well ye knew those amongst you who transgressed in the matter of the Sabbath: We said to them: "Be ye apes, despised and rejected." (Sura 2: 66)[25]

> Say: "Shall I point out to you something much worse than this, (as judged) by the treatment it received from Allah? Those who incurred the curse of Allah and His wrath, those of whom some He transformed into apes and swine, those who worshipped evil; these are (many times) worse in rank, and far more astray from the even path!" (Sura 5:60)

Unlike these verses from the Koran, references to dehumanization in the *hadith* compiled two or three centuries later have a distinctly anti-Jewish flavor. They describe how a group of Israelites were transformed into rats, how unbelievers are turned into monkeys and pigs, and how Abraham's father was transformed into an animal and hurled into the raging fires of hell.[26]

Whereas both Boethius and Aristotle entertained the idea that one could be outwardly human but inwardly subhuman, there's no trace of this idea in the Muslim texts dating from this period. The zoomorphic transformations described in the Koran and *hadith* were apparently assumed to be concrete, physical metamorphoses: a person's body was transformed into that of an ape, a pig, or a rat. But when Aristotelian philosophy began to make serious inroads into Islamic thought during the ninth century, Muslim philosophers started to interpret these stories metaphorically, as references to the bestial degradation of the human soul. During this period they also began to speculate about the ranks occupied by different sorts of people on the great chain of being, just as their European counterparts would do centuries later. Some groups (in particular, sub-Saharan Africans) were singled out as verging on the subhuman. I will discuss this further in Chapter Four.[27]

Back in Europe, the Boethian theory persisted right up to the end of the Renaissance. The famous "Oration on the Dignity of Man" written in 1486 by the Renaissance scholar Giovanni Pico della Mirandola, begins with a discussion of our rank on the great chain of being, and goes on to talk about the human potential for reverting to a less than human condition.

> What is this rank on the chain of being? God the Father, Supreme Architect of the Universe, built this home, this universe we see all around us, a venerable temple of his godhead, through the sublime laws of his ineffable Mind. The expanse above the heavens he decorated with Intelligences, the spheres of heaven with living, eternal souls. The scabrous and dirty lower worlds he filled with animals of every kind.

God created human beings because he desired "some creature to think on the plan of his great work, and love its infinite beauty, and stand in awe at its immenseness." But He couldn't find a prototype for their design because creation was already complete. One might imagine that an omnipotent deity could get around this problem by conjuring up a new level out of nothing and inserting it in an appropriate position on the chain, but Pico tells us that He dealt with it by creating humans with no fixed nature—that is, as beings capable of choosing their own nature.

> [T]he Great Artisan . . . made man a creature of indeterminate and indifferent nature, and, placing him in the middle of the world, said to him "Adam, we give you no fixed place to live, no form that is peculiar to you, nor any function that is yours alone. According to your desires and judgment, you will have and possess whatever place to live, whatever form, and whatever functions you yourself choose. All other things have a limited and fixed nature prescribed and bounded by Our laws. You, with no limit or no bound, may choose for yourself the limits and bounds of your nature. . . .

Pico believed that we are free to choose our position on the great chain, and that the ability to determine one's own destiny is a uniquely human attribute. Some people opt to ascend toward God, while others sink to the level of beasts. "To you is granted the power of degrading yourself into the lower forms of life, the beasts," wrote Pico, "and to you is granted the power, contained in your intellect and judgment, to be reborn into the higher forms, the divine."[28]

Pico's near contemporary, the larger-than-life Swiss physician Philippus Aureolus Theophrastus Bombastus von Hohenheim (or Paracelsus, as he is conveniently called) also wrote about dehumanization. Whereas Pico believed that human beings freely choose their essence, Paracelsus held that human and subhuman elements strive *within* our nature. The idea that the animal and divine essences wrestle for dominance in the human soul was fairly standard theological fare during the medieval period. We find it, for instance, in Saint Gregory the Great's *Homily of the Gospel*, as well as in the writings of Irish theologian Johannes Scotus

Eriugena, both of whom informed their readers that human beings are composed of every creature."[29] Paracelsus embraced a typically idiosyncratic version of the animal-in-man doctrine. In the words of historian of science William R. Newman:

> Man has both a spiritual and an animal capacity and that when one calls a man a wolf or dog, this is not a matter of simile but of identity. . . . When someone acts in a bestial fashion, he therefore actualizes the beast within and literally becomes the animal whose behavior he imitates.[30]

This theory had obvious affinities with those expounded by Boethius and Pico, but its emphasis was slightly different. For Paracelsus, we become dehumanized by virtue of actualizing a subhuman potential, something that was in us all along, in a latent or suppressed state. By yielding to the beast within, we forego our humanity.

From the medieval perspective, dehumanization was viewed as an actual transformation from a human to a subhuman state, caused by sinful behavior. All this would change with the dawning of the Enlightenment.

THE ENLIGHTENMENT: DAVID HUME AND IMMANUEL KANT

> Many of the Christians do not esteem, not look at us any otherwise, than Dogs. . . . We are not Beasts as you count and use us, but rational souls.
>   —THOMAS TRYON, *FRIENDLY ADVICE TO THE*
> *GENTLEMEN-PLANTERS OF THE EAST AND WEST INDIES*[31]

It wasn't until over a thousand years after Boethius's death that anyone made a start at developing a psychological account of dehumanization. The first person to take up the challenge seems to have been the Scottish philosopher David Hume.

Hume was a child of the Enlightenment. Born in 1711, he was expected to follow in his father's footsteps and become an attorney. To this end, he was bundled off to the prestigious University of Edinburgh at the age of twelve, which was precocious even by eighteenth-century standards (the normal university entrance age was fourteen). But Hume despised the study of law. He confessed in an autobiographical sketch that the prospect of becoming a lawyer made him "nauseous," and that even as a child he was drawn to the life of a scholar and philosopher. After leaving school, and an ill-fated stint as a clerk for a sugar-importing business in Bristol, he suffered a nervous breakdown and retreated to the town of La Flèche in western France for rest and recuperation. While there, Hume wrote his masterpiece, a book that was poorly received during his lifetime but now stands as one of the greatest philosophical works in the English language. It was published in two installments in 1739 and 1740 under the title A *Treatise of Human Nature*.[32]

The philosophical thrust of the *Treatise* was to bring scientific methods to bear on the study of human nature. Science had made tremendous strides in the century or so before his birth, and the momentum of discovery was, if anything, accelerating during his lifetime. There were radical developments in physics, culminating in the publication of Isaac Newton's *Principia* in 1687, which explained the variegated phenomenology of the physical world with breathtaking economy and precision (Newtonian theory was so formidable that it was not substantially improved until Albert Einstein came on the scene at the dawn of the twentieth century). Although there were as yet no methods for investigating the microstructure of the physical world, seventeenth-century physicists—or "natural philosophers" as they were then called—speculated that all material objects are composed of miniscule "corpuscles," the character and arrangement of which determines all of their observable properties, and thus laid the conceptual foundations for particle physics. Practitioners of the new science of chemistry, which had only recently emerged from the hocus-pocus of alchemy, were beginning to understand the principles by which physical substances interact with one another. Advances in optics opened up hitherto undreamed of vistas, from microscopic forms of life to the distant moons of Jupiter. Knowledge was expanding at a vertiginous pace.

Not every discipline was swept along by this torrent of scientific progress. Psychology remained a backwater, virtually stagnant since the days of the ancient Greeks. While still in his teens, Hume recognized that the extraordinary power of science lay in its method, and was convinced that the scientific method is our only hope for unlocking the mysteries of human nature. In Hume's day, psychology was based on claims that derived their authority either from Christian orthodoxy and the time-honored pronouncements of Plato and Aristotle (especially as viewed through the lens of medieval Christian thinkers), or else from speculative thought uninformed by empirical observation.

Hume thought that this approach was wrongheaded, and argued that psychology ought to free itself from both ancient tradition and armchair speculation. It should "reject every system . . . however subtle or ingenious, which is not founded on fact and observation," and "hearken to no arguments but those which are derived from experience."[33]

Hume's earliest reference to dehumanization is found in a passage from the *Treatise*. His tantalizingly brief comments appear in a discussion of the origins of love and hate. Hume begins by pointing out that we tend to feel affection toward people who give us pleasure, and aversion toward people who cause us displeasure. "Nothing is more evident," Hume assures us, in his delightful eighteenth-century English, "than that any person acquires our kindness, or is expos'd to our ill-will, in proportion to the pleasure or uneasiness receiv'd from him, and that the passions keep pace exactly with the sensations in all their changes and variations." Having made this point, he focuses on a special case of "uneasiness receiv'd."

> When our own nation is at war with any other, we detest them under the character of cruel, perfidious, unjust and violent: but always esteem ourselves and allies equitable, moderate and merciful. If the general of our enemies be successful, 'tis with difficulty we allow him the figure and character of a man. He is a sorcerer: he has a communication with daemons . . . he is bloody-minded and takes a

pleasure in death and destruction. But if the success be on our side, our commander has all the opposite good qualities, and is a pattern of virtue, as well as of courage and conduct. His treachery we call policy: His cruelty is an evil inseparable from war. In short, every one of his faults we either endeavor to extenuate, or dignify it with the name of that virtue, which approaches it. 'Tis evident that the same method of thinking runs thro' common life.[34]

This is an elegant description of what present-day social psychologists call *outgroup bias*—the tendency to favor members of one's own community and discriminate against outsiders (otherwise known as the "us and them" mentality). *We* are more industrious, conscientious, attractive, and so on, than *they* are. When things go badly for one of *us* it's an injustice, but when the same thing happens to one of *them* it's because they brought it on themselves. On the flip side, when one of *us* experiences good fortune, it's richly deserved, but when one of *them* benefits from a windfall it's an undeserved stroke of good luck.

As the quoted passage shows, Hume thought that this principle applies to international relations just as much as it does to everyday life. And he was right. Jerome D. Frank's 1982 essay "Prenuclear-Age Leaders and the Nuclear Arms Race" lets the facts speak for themselves.

Every so often [Gallup polls] ask their respondees to select from a list of adjectives the ten which best describe members of other nations. . . . Back in 1942, Germany and Japan were our bitter enemies, and Russia was our ally; and in 1942, among the first five adjectives chosen to characterize both the Germans and the Japanese were: "warlike," "treacherous," and "cruel." None of these appeared in the list for the Russians at that time. In 1966, when Gallop surveyed responses to mainland China, the Chinese were seen as "warlike," "treacherous," and, being Orientals, "sly." After President [Richard] Nixon's visit to China, however, these adjectives disappeared about the Chinese, and they [were] . . . characterized as "hard-working," "intelligent," "artistic," "progressive," and "practical."[35]

Hume takes the idea of outgroup bias even further by arguing that sometimes we are so strongly biased against others that we stop seeing them as human beings. He may have had in mind here a remark by his predecessor John Locke, who stated in his *Two Treatises of Government* (published in 1690, and certainly familiar to Hume) that tyrants "may be destroyed as a lion or tiger, one of those wild savage beasts with whom one can have no society or security." (This passage is often incorrectly interpreted as a reference to the destruction of American Indians. In fact, Locke probably had the English despot Charles II in mind.) Whatever his source of inspiration, present-day studies of wartime propaganda confirm Hume's observation. As we have already seen in Chapter One, wartime leaders are often portrayed as dangerous, nonhuman creatures—predators, demons, monsters, and the like. Oddly, Hume doesn't claim that we perceive enemy leaders as subhuman. His examples seem to imply that we imagine them as diabolically superhuman.[36]

Sketchy though these comments are, we can see in them the first glimmerings of a psychological theory of dehumanization.

Hume picked up the thread again in his 1751 *An Enquiry Concerning the Principles of Morals*. As its title suggests, *Enquiry* is an analysis of the nature of moral judgment. What exactly goes on when we consider an act to be good or bad, right or wrong? Hume rejected the view that our moral attitudes are based on rules or principles. Instead, he believed that morality is primarily a matter of feeling. More specifically, he argued that our moral judgments flow from a psychological faculty that he called *sympathy*.

The word *sympathy* meant something quite different during the eighteenth century than it does today. In Hume's writings (as well as those who were influenced by him, such as the economist Adam Smith) *sympathy* doesn't mean commiseration. Rather, it refers to an inborn tendency to resonate with others' feelings—to suffer from their sorrows and to be uplifted by their joys (as he colorfully put it in the *Treatise*, "the minds of men are mirrors to one another"). Our resonance with others evokes feelings of approval or disapproval in us, and these feelings are the basis of our moral judgments. Suppose you were to witness an adult cruelly beating a child, and feel a sense of moral outrage welling up in you. Hume would say that you are "tuning in" to the felt quality of the child's experience, and that the feeling of moral disapproval is produced by the hurt evoked in you by witnessing the scene.

Our sympathies tend to be unevenly distributed. We care much more for certain people than we do for others. The odds are that a loved one suffering from the flu evokes vastly more sympathetic concern in you than the fact that millions of people live in abject poverty. Hume was well aware of how our sympathies become skewed, and he identified three powerful sources of bias, arguing that we naturally favor people who resemble us, who are related to us, or who are nearby. The people who are "different"—who are another color, or who speak a different language, or who practice a different religion—people who are not our blood relations or who live far away, are unlikely to spontaneously arouse the same degree of concern in you as members of your family or immediate community. But morality can't play favorites—it has to apply across the board. So, we must bring our biases to heel by adopting what Hume called the "common point of view." This, too, is accomplished by the power of imagination. We try to detach ourselves from our own limited perspective and, putting ourselves in other people's shoes, imagine how they would respond to the situation. The resulting feeling of approval or disapproval fixes our moral verdict.

So, what's the connection between sympathy and justice? First of all, it's important to understand that when Hume talks about justice, he has quite a narrow purview. He's concerned with property rights pure and simple, rather than justice in the broader sense of the word. He argues that the demand for justice arises from two facts—one about the world and the other about human nature. The first is that resources are limited: Many of the things that we need and want come in finite quantities, and this makes it impossible for everyone to have as much as they desire or require. The second is selfishness—the raw fact that we tend to favor ourselves over others, and our friends and family over strangers. Greed, writes Hume, is an "insatiable, perpetual, universal" socially destructive feature of human nature. Chaos would reign if we all were left to our own devices, so we adopt rules that "bestow stability on the possession . . . of external goods, and leave every one in the peaceable enjoyment of what he may acquire by his fortune and industry." For in this way "everyone knows what he may safely possess; and the passions are restrained in their partial and contradictory motions."

Justice becomes a moral issue, rather than simply a pragmatic one, when we take the common point of view and allow sympathy to go about its work. Morality enters the picture to the extent that we are able to emotionally resonate with the circumstances of others, and thereby experience their property rights as though they were our own.[37]

Having set out his theory of justice, Hume invites his readers to imagine that there is "a species of creatures, intermingled with men, which, though rational, were possessed of such inferior strength, both of body and mind, that they were incapable of all resistance, and could never, upon the highest provocation, make us feel the effect of their resentment. . . ." If this were the case, Hume asks, how would we behave toward these animals? His answer is disturbing. Although we might "give gentle usage to these creatures," he says, we would not "lie under any restraint of justice with regard to them, nor could they possess any right or property."

> Our intercourse with them could not be called society, which supposes a degree of equality; but absolute command on the one side, and servile obedience on the other. Whatever we covet, they must instantly resign: Our permission is the only tenure by which they hold their possessions: Our compassion and kindness the only check, by which they curb our lawless will: and as no inconvenience ever results from the exercise of a power, so firmly established in nature, the restraints of justice and property, being totally *useless*, would never have place in so unequal a confederacy.[38]

The thrust of Hume's reasoning is clear. The purpose of justice is to ensure social harmony. But the nonhuman creatures described in

his conceit do not have the wherewithal to participate in human society, and this is why considerations of justice are irrelevant to our deal-
ings with them. Hume next goes on to point out that this isn't just a
thought-experiment, but is "plainly the situation of men, with regard to
animals." Because nonhuman animals cannot participate in human
society, the notion of justice is inapplicable to them. It is at this point
that dehumanization enters the picture. Hume points out, "The great
superiority of civilized Europeans, above barbarous Indians, *tempted us
to imagine ourselves on the same footing with regard to them, and made
us throw off all restraints of justice, and even of humanity, in our treatment of them.*"[39]

To make sense of this passage, we need to do some reading between
the lines. The key is Hume's view of the intimate connection between sympathy and imagination. Hume held that it is only because we
can imagine that others are beings like us that we can "enter . . . into
the opinions and affections of others, whenever we discover them."

Think about this for a moment. As you go through your daily life,
you interact with other people, and as you do so, you assume that they
are conscious beings with beliefs and desires much like those with which
you are acquainted. But what evidence do you have for this? Couldn't it
be that you are the only person in the world with a subjective mental
life, while everyone else—including your nearest and dearest—are zombies or fancy automata? That possibility seems absurd; at best, the stuff
of schizophrenia or science fiction. Although we don't presume to
know the intimate details of other people's mental states, we are confi-
dent that their experiences, thoughts, feelings, and so on, are more or
less the same as our own.

But what's the source of our confidence? What entitles us to think
that others have inner lives? We can't *perceive* other people's experiences, because they are beyond the range of our sense organs. Neither
can we vouch for their existence through a process of logical inference.
In Hume's view, attributing mental states to others is the work of the
imagination. We recognize that others are outwardly similar to us, and
then take the imaginative leap of attributing mental states to them that
are broadly similar to our own. This doesn't have to be a deliberate

process. It's more plausibly seen as automatic. And once it happens, the stage is set for sympathy to kick in.[40]

However, imagination doesn't always get things right. In this connection, there are two ways that it can err. One way is by producing anthropomorphic illusions. Hume remarked, "There is a universal tendency among mankind to conceive all beings like themselves, and to transfer to every object, those qualities, with which they are familiarly acquainted, and of which they are intimately conscious."

> We find human faces in the moon, armies in the clouds;
> and by a natural propensity, if not corrected by experience
> and reflection, ascribe malice or good-will to every thing,
> that hurts or pleases us. Hence . . . in poetry . . . trees,
> mountains, and streams are personified, and the inanimate
> parts of nature acquire sentiment and passion.[41]

The second source of error is when imagination tricks us into believing that others aren't really human, and thereby prevents us from sympathizing with them. Although Hume never propounds this explicitly, his brief discussion of European colonialism seems to imply it. As I understand him, Hume thought that disordered imagination interfered with the colonists' ability to sympathize with Native Americans. They imagined Indians to be an alien species, and this dissolved their moral restraint.

Hume completes the discussion of colonialism with some remarks on the oppression of women by observing, "In many nations, the female sex are reduced to like slavery, and are rendered incapable of all property, in opposition of their lordly masters." Women are deprived of justice, but not, it seems, of humanity.[42]

David Hume wasn't the only intellectual of his era to condemn the brutalities of colonialism, nor was he unique in recognizing that dehumanization played a role in it. His friend, the French philosopher Denis Diderot, was far more outspoken. "Savage Europeans!" he raged. "You doubted at first whether the inhabitants of the regions you had just discovered were not animals which you might slay without remorse because they were black and you were white. . . . In order to repeople one part of the globe you have laid waste, you corrupt and depopulate another."[43] The difference between Hume and figures like Diderot is that although Hume condemned colonialism, he wasn't content *just* to condemn it. He tried to identify the psychological processes responsible for the colonists' behavior, and to situate this explanation in a comprehensive theory of human nature.

The next important contributor to the concept of dehumanization was Immanuel Kant, who was arguably the last of the great Enlightenment thinkers. Born in 1724, Kant was an intellectually conservative German academic until he began to read Hume at some point in his forties. This experience opened his eyes and, as he famously put it, "interrupted my dogmatic slumbers."[44]

Kant set out to extinguish the skeptical fires set by Hume, fires that threatened to engulf and consume cherished Enlightenment beliefs about the sovereignty of reason. Hume had cast doubt on a whole slew of philosophical beliefs, including the idea of cause and effect, the integrity of inductive reasoning, the belief in an inner self, and—perhaps most troubling for Kant—beliefs about the foundations of moral judgment. If morality is just a matter of how we feel, then moral values seem to lose all of their objectivity.

Kant's approach to morality is, in some respects, the very antithesis of Hume's. Whereas Hume's theory was based on feeling, Kant's was based on reason. Kantian theory makes a sharp distinction between *means* and *ends*. When we value something as a means, we treat is as a stepping-stone to some further goal. For example, as I type these words, I'm sipping from a mug of strong black coffee. I value the coffee because it will help me stay alert while I am writing (it's early evening, and I'm planning to work on this manuscript far into the night). I'm not drinking the coffee because drinking coffee is intrinsically right. I'm drinking it as a stepping-stone to achieving something else.

Sometimes, we use other people as means to ends. My students use me as a means of getting an education, and I use them as a means of earning a living. But we don't relate to one another *purely* as means. I also value them as human beings, as they—I trust—value me. If we were to treat one another only as means, our relationship would be mutually exploitative rather than respectful. These sorts of considerations are central to Kant's vision of an ethical life. He argued that we are duty-bound to "treat humanity, whether in your own person or in the person of any other, always at the same time as an end, never merely as a means."[45]

Kant didn't extend this principle to our relationships with nonhuman animals. He thought that human beings have absolute worth—that is, value in and of themselves—and that, in this respect, people are "altogether different in rank and dignity from things, such as irrational animals, with which one may deal and dispose at one's discretion." Unlike Hume, Kant believed that nonhuman animals "have only a relative worth, as means, and are therefore called things, whereas rational beings are called persons because their nature . . . marks them out as an end in itself."[46] Animals have neither property rights nor any moral standing.

> When he first said to the sheep, "the pelt which you wear was given to you by nature not for your own use, but for mine" and took it from the sheep to wear it himself, he became aware of a prerogative which, by his nature, he enjoyed over all the animals; and he now no longer regarded them as fellow creatures, but as means and instruments to be used at will for the attainment of whatever ends he pleased.

Kant thought that this has an important implication for our relationship with fellow human beings.

> This . . . implies . . . an awareness of the following distinction: man should not address other human beings in the same way as animals, but should regard them as having an equal share in the gifts of nature.[47]

Although he never discussed dehumanization as such, Kant recognized that people are prone to regard one another only as means. When we do this, we place others in the same category as subhuman creatures and thereby exclude them from the universe of moral obligation. It then becomes morally permissible to "deal and dispose" of them as we please.

THE RISE OF ANTHROPOLOGY: WILLIAM GRAHAM SUMNER

A lot of guys really supported the whole concept that if they don't speak English and they have darker skin, they're not as human as us, so we can do what we want.
— JOSH MIDDLETON, US ARMY 82ND
AIRBORNE DIVISION[48]

As the European powers consolidated their grip on Africa, Australasia, and North America during the eighteenth and nineteenth centuries, settlers, missionaries, and explorers increasingly encountered cultures that were very different from their own. The relatively narrow domain of European scholarship was challenged and enriched by exposure to alien ways of life, and by the middle of the nineteenth century the new discipline of social anthropology had emerged from the mix.

The early anthropologists collected the strange and occasionally hair-raising stories brought back from overseas by missionaries, explorers, and soldiers of fortune, and used them to craft theories about human nature and the evolution of culture. Of course, no science deserving of the name can be based on anecdotes, and anthropologists began to realize around the turn of the twentieth century that they needed more rigorous methods for gathering data. When they eventually got up from their plush Victorian armchairs and started observing cultures firsthand, they noticed that people everywhere tended to think of their own culture as superior to everyone else's.

In 1907, Yale University sociologist William Graham Sumner gave this tribal tendency a name; he called it *ethnocentrism*. As Sumner described it, ethnocentrism is the belief that "one's own group is the center of everything, and all others are scaled and rated with reference to it . . . Each group nourishes its own pride and vanity, boasts itself superior, exalts its own divinities, and looks with contempt on outsiders."[49] As ethnographic research grew more extensive, Sumner's claims were amply confirmed.[50]

The most extravagant expression of ethnocentrism is the belief that the members of one's own culture are the only true human beings, and it's at this point that ethnocentrism begins to shade into dehumanization. Sumner illustrates the point with a number of examples.

> When the Caribs were asked whence they came, they answered "We alone are people." The meaning of the name Kiowa is "real or principal people." The Lapps call themselves "men" or "human beings." . . . The Tunguses call themselves "men." As a rule, it is found that native peoples call themselves "men." Others are something else—perhaps not defined—but not real men. In myths, the origin of their own tribe is that of the real human race. They do not account for the others.[51]

Over three decades later, Franz Boas, who is credited as the founder of modern cultural anthropology, observed that, "Among many primitive people, the only individuals dignified by the term human beings are members of the tribe. It even happens in some cases that language will designate only tribal members as 'he' or 'she,' while all foreigners are 'it' like animals."[52] A quick survey of Native American tribal names drives the point home. Many Native American tribes (including the Inuit, Tanaina, Chipewyan, Navajo, Kutchin, Innu, Klamath, Apache, Mandan, Comanche, Ute, Hurok, and Cheyenne) refer to themselves as "the human beings" (as do contemporary Germans—the word *Deutsch* comes from an Indo-European root meaning "human beings").

Nowadays, the word *ethnocentrism* is used to express moral disapproval. Accusations of ethnocentrism are almost always used to disparage blinkered Western views of indigenous cultures. But Sumner used it in a descriptive way rather than as an evaluation. Of course, it's true that Westerners often display ethnocentric biases toward aboriginal people—but, as we have seen, it's equally true that aboriginal communities often think of Westerners and members of other aboriginal groups as less than human.

Napoleon Chagnon, an anthropologist best known for his studies of the Yanomamö of Brazil and Venezuela (and who will return in Chapter Seven), gives an engaging account of his experience at the receiving end of tribal ethnocentrism. When he first made contact with the Yanomamö, "They were pushy, they regarded me as subhuman or inhuman, they treated me very badly." But eventually:

> More and more of them began to regard me as less of a foreigner or a sub-human person and I became more and more like a real person to them, part of their society. Eventually they began telling me, almost as though it were an admission on their part: "You are almost a human being, you are almost a Yanomamö."[53]

Sumner believed that ethnocentrism was found the world over—in modern nation-states as well as in primitive tribes. "Each state now regards itself as the leader of civilization," he wrote, "the best, the freest, and the wisest, and all others as inferior. . . . The patriotic bias is a recognized perversion of thought and judgment against which our education should protect us."[54] Less than a decade after he wrote these words, a frenzy of patriotic bias engulfed Europe, inundating the continent in blood. World War I took slaughter to an unprecedented level. It left around 17 million dead (about a million of whom perished of starvation) and many millions maimed or seriously injured. This human cataclysm led thoughtful people to ask searching questions about war and human nature.

THE GREAT WAR: JOHN T. MACCURDY

> What passing bells for these who die as cattle?
> Only the monstrous anger of the guns
> —WILFRED OWEN, "ANTHEM FOR DOOMED YOUTH"[55]

One of the most significant attempts to address these questions was a slim volume called *The Psychology of War*, published in 1918 by a neurologist named John T. MacCurdy. MacCurdy had an unusual trajectory. Born in 1886 into an academic family (his father was professor of

Assyriology at the University of Toronto), he first studied biology at Toronto and then took a medical degree at Johns Hopkins. Sometime after completing his studies, he met Sigmund Freud's English-speaking disciple Ernest Jones, and became one of the founding members of the American Psychoanalytic Association and its president. The final stage of his career was spent at Cambridge University, where he was lecturer (in American parlance, "professor") of psychopathology until his death in 1947.[56]

When the United States entered World War I in 1917, MacCurdy became a member of the American Expeditionary Force, and visited hospitals in the United Kingdom where shell-shocked soldiers received psychiatric treatment. This experience led him to think deeply about the psychological dynamics of war. MacCurdy was struck by the degree to which warfare depends on group solidarity. People live in groups cemented by powerful ties of community loyalty, and this collective devotion to the group makes mass violence possible.

> Here we have what is perhaps the greatest paradox of human nature. The forgetting of self in devotion to others, altruism or loyalty, is the essence of virtue. At the same time, precisely the same type of loyalty that makes of a man a benefactor to all mankind can become the direst menace to mankind when focused on a small group.[57]

Altruism and group loyalty are necessary for war, but they're not sufficient. War can't occur unless the members of one group are prepared to go out and kill the members of another. This raises a problem. In the movies, killing is easy. Both heroes and villains nonchalantly blow their enemies away, unperturbed by hesitation or remorse. But in real life things are different. Unless one is a sociopath, a psychologically disturbed person devoid of empathy and moral feeling, there are strong inhibitions against killing others.

Unlike many writers on war and violence, MacCurdy was acutely aware of our inhibitions against killing, and he pointed out that "unless the animosity of the race becomes individual, it would be impossible for a civilized man to deal a lethal blow, restrained as he is by the inhibitions of generations."[58] How, then, do warriors overcome their ingrained resistance to killing other humans? MacCurdy thought that our ability to dehumanize others is part of the answer. He framed his argument by painting a picture of prehistoric tribes vying for scarce resources.

> In earlier days . . . friction with other tribes over hunting grounds or other coveted possessions must have made strangers appear like those of other species. . . . Advance of knowledge has taught that all the members of the species Homo sapiens are men, but it is doubtful whether that knowledge is a vital part of our automatic mental life. It is one thing for us to recognize in an animal, identity of anatomical structure, and another to feel that he is like ourselves. Without this instinctive bond, every stranger, every member of every other group, must to a greater or less extent arouse in us the biological reaction appropriate towards a different species. We have sympathy for a dog, an animal useful to us, but we kill wolves, snakes and insects without any revulsion of feeling for the act.[59]

Although we now know that all people are members of the same species, this awareness doesn't run very deep, and we have a strong unconscious ("automatic") tendency to think of foreigners as subhuman creatures. This gut-level assessment often calls the shots for our feelings and behavior. We can bring ourselves to kill foreigners because, deep down, we don't believe that they are human. MacCurdy emphasizes that these beliefs are not always unconscious. When tensions are high, "[t]he unconscious idea that the foreigner belongs to a rival species becomes a conscious belief that he is a pestiferous type of animal."[60]

This was the first full-blown psychological theory of dehumanization, but it was for the most part ignored. After 1918, the study of dehumanization was neglected for the better part of half a century, until a psychoanalyst named Erik Erikson introduced the concept of cultural pseudospeciation.

PSYCHOANALYSIS: ERIK H. ERIKSON

And mercy on our uniform,
Man of peace or man of war,
The peacock spreads his fan.
                    —LEONARD COHEN, "THE STORY OF ISAAC"[61]

Erik Homberger Erikson led an unlikely life. Born in Germany in 1902, the child of his Danish mother's extramarital affair, Erikson's formal education ended with high school. After that, he took to the road, hitchhiking across Europe and eking out a living as an itinerant artist. At the age of twenty-five, he drifted into Vienna and found a summer job as an elementary school teacher. The Hitzig School, where the young Erikson found himself, was not an ordinary one. It was a liberal, experimental school founded by Sigmund Freud's daughter Anna and her longtime companion Dorothy Tiffany Burlingham. Erikson flourished in the vaguely bohemian ambience of both the school and Viennese psychoanalytic scene, and with Anna's encouragement and support he remained in Vienna and become a psychoanalyst.

Five years later, Hitler was sworn in as chancellor of Germany, and many members of the predominantly Jewish psychoanalytic movement saw the writing on the wall. They fled Austria and Germany, initially to France, Belgium and Scandinavia, and later, as the Nazi shadow lengthened, to Latin America, Great Britain, and the United States. Erikson was part of this diaspora. He arrived in New York City in 1933, and became an American citizen in 1938 (the same year that Austrians lined the streets welcoming German troops into Vienna with shouts of "Heil Hitler!" and the elderly Sigmund Freud fled to London to die, as he put it, "in freedom"). Erikson's move to America marked the start of his meteoric rise to intellectual celebrity. After a series of influential publications, and still with only a high school diploma behind him, he was offered a special professorship at Harvard University, where he remained until his retirement at the age of sixty-eight.

The question of how culture shapes identity was the axis around which Erikson's work revolved. He was fascinated by the fact that although all human beings are all members of the same species, we tend to treat the members of different cultures as different kinds of beings. This invites a comparison with biological taxonomy. Just as biological lineages bifurcate to form separate species, human populations coalesce into separate cultures. Human cultures are artificial species, or, more accurately, pseudospecies.

Erikson introduced the term *pseudospecies* in 1966, at a meeting of the Royal Society of London. The famous Austrian biologist Konrad Lorenz was in the audience, and suggested that he use the term *cultural pseudospeciation* to describe the process of cultural differentiation. Erikson adopted Lorenz's advice. The term caught on, and proliferated rapidly through the social science literature.

Erikson wrote surprisingly little on pseudospeciation. It's mentioned in passing in several of his writings, but he wrote only one short paper specifically devoted to it, and that paper is less than four pages long.[62] Here's how Erikson defined it.

> The term denotes that while man is obviously one species, he appears and continues on the scene split up into groups (from tribes to nations, from castes to classes, from religions to ideologies and, I might add, professional associations) which provide their members with a firm sense of unique and superior human identity—and some sense of immortality.[63]

Ritualistic paraphernalia such as "pelts, feathers, and paints, and eventually costumes and uniforms" as well as "tools and weapons, roles and rules, legends, myths and rituals" confirm and reinforce these cultural identities. But pseudospeciation is not all sweetness and light. It's also the basis for mass violence and oppression.

What has rendered this . . . process a potential malignancy of universal dimensions, however, is that, in times of threatening technological and political change and sudden upheaval, the idea of being the preordained foremost species tends to be reinforced by a fanatic fear and anxious hate of other pseudospecies. It then becomes a periodic and often reciprocal obsession of man that these others must be annihilated or kept "in their places" by periodic warfare. . . . [64]

It's important to be clear about what Erikson was and wasn't saying. There are two points to be made in this connection. First, although he was not always consistent, Erikson intended *pseudospeciation* as a descriptive term—an engaging metaphor for the tendency of our species to coalesce into diverse, mutually exclusive social groups with ethnocentric biases. He didn't think that pseudospeciation explained anything. Perhaps an example will make this a bit clearer. *Affluence* is just a word for having lots of money. Clearly, it would be uninformative to say of Bill Gates that he is affluent because he has lots of money. That would be equivalent to saying that Bill Gates has lots of money because he has lots of money! It's uninformative to say cultures are formed *because* of cultural pseudospeciation for exactly the same reason. Second, although Erikson sometimes mentioned dehumanization in discussions of cultural pseudospeciation, he didn't equate the two. He thought of pseudospeciation as necessary but not sufficient for dehumanization.

These distinctions were quickly eroded as the term gained wider currency, as is evidenced by the following passage by Konrad Lorenz, who managed to commit both errors in the space of a single paragraph.

The dark side of pseudospeciation is that it makes us consider the members of pseudospecies other than our own as not human, as many primitive tribes are demonstrably doing, in whose language the word for their own particular tribe is synonymous with "Man." From their viewpoint it is not strictly speaking cannibalism if they eat fallen warriors of an enemy tribe.[65]

Lorenz knew whereof he spoke. During the 1930s, he had been an active member of the Nazi party, and endorsed their racial policies. For example, he wrote in 1940 in the German periodical *Der Biologe* that:

> There is a certain similarity between the measures which need to be taken when we draw a broad biological analogy between bodies and malignant tumors on the one hand and a nation and individuals within it who have become asocial because of their defective constitution, on the other hand. . . . Fortunately, the elimination of such elements is easier for the public health physician and less dangerous for the supra-individual organism, than such an operation by a surgeon would be for the individual organism.[66]

Erikson never even gestured toward an explanation of why cultural pseudospeciation occurs. Explaining it as a by-product of culture would put the cart before the horse: Cultural diversity is supposed to be the outcome of pseudospeciation, so it can't also be its cause. Clearly, if we want to discover why cultures form we must look in the precultural domain for an explanation. To do this, we need to turn to biology, and consider the evolutionary forces that shaped the human animal.

BIOLOGICAL ROOTS: LORENZ, EIBL-EIBESFELDT, AND GOODALL

> On the outer door of a mall in my town is a sign: NO ANIMALS ALLOWED. Look inside, however, and one sees dozens of human animals, browsing through clothing racks or standing behind cash registers. On what do people base their certainty that the term *animals* does not apply to them?
>
> —MARIAN SCHOLTMEIJER, *WHAT IS HUMAN?*

When Erikson introduced the concept of pseudospeciation in 1966, there were a number of prominent biologists present, including Konrad

Lorenz, who took up the concept and included a short discussion of it in his 1966 book *On Aggression*. But Lorenz was not the only biologist who found the concept of pseudospeciation useful. During the late 1960s and early 1970s there was growing interest in biological explanations of human behavior. In Austria, the new field of human ethology was being pioneered by Lorenz's colleague Irenäus Eibl-Eibesfeldt, who used principles derived from the study of animal behavior to explain the behavior of *Homo sapiens*. At the same time, a small group of Harvard biologists spearheaded by E. O. Wilson was developing the new discipline of sociobiology—the biological study of social behavior, including human social behavior. Scientists from both of these groups adopted the concept of cultural pseudospeciation.

Perhaps the best way to explain why biologists were attracted to the notion of pseudospeciation is to start with some reflections on culture. Many people, both inside and outside the academy, assume that culture is what sets human beings off from the rest of the animal kingdom. It is common—especially in the humanities—to dichotomize biology and culture, and to assume that this binary opposition defines an unbridgeable gulf between human beings and other animals. But nature dislikes yawning dichotomies. It is continuous rather than discrete, preferring subtle gradations to abrupt discontinuities. Perhaps, then, behavior of nonhuman animals can disclose the driving forces behind human culture.

Many animals spend their lives in close-knit communities. Sometimes individuals discover novel forms of behavior, which group members copy. If these prove useful, they may be transmitted down the generations—not by genetic inheritance, but by custom. These traditions are probably continuous with (although obviously far more rudimentary than) human culture.

Chimpanzees are especially adept at creating and sustaining traditions, which makes for striking differences between local populations. More that forty populations of chimpanzees across the breadth of sub-Saharan Africa have been found to use tools, and each of them uses tools in ways that are different from the rest.[67] As Harvard primatologist Richard Wrangham and science journalist Dale Peterson point out,

"Chimpanzee traditions ebb and flow, from community to community, across the continent of Africa."

> On any day of the year, somewhere chimpanzees are fishing for termites with stems gently wiggling into curling holes, or squeezing a wad of chewed leaves to get a quarter cup of water from a narrow hole high up in a tree. Some will be gathering honey with a simple stick from a bee's nest, while others are collecting ants by luring them onto a peeled wand, then swiping them into their mouths. There are chimpanzees in one place who protect themselves against thorny branches by sitting on leaf-cushions, and by using leafy sticks to act as sandals and gloves. Elsewhere are chimpanzees who traditionally drink by scooping water into a leaf cup, and who use a leaf as a plate for food. There are chimpanzees using bone picks to extract the last remnants of the marrow from a monkey bone, others digging with stout sticks into mounds of ants or termites, and still others using leaf napkins to clean themselves or their babies. These are all local traditions, that have somehow been learned, caught on, spread, and been passed across generations among apes living in one community or a local group of communities but not beyond.[68]

One force that's instrumental in maintaining this sort of cultural diversity is hostility between communities. Although there are some exceptions, social mammals tend to be fiercely xenophobic. Aggression between community members is usually muted, but unbridled violence is readily unleashed against outsiders who have the misfortune of being in the wrong place at the wrong time. Violence is also meted out to deviant members within a community—individuals who violate group norms of appearance and behavior become targets of community aggression. Mark a hen's comb with an oddly colored spot, or tie it so it hangs in a peculiar direction, and her former flock mates will attack her mercilessly. Jane Goodall, who was the first scientist to observe chimpanzees up close and personal in the wild, noticed that crippled chimpanzees were rejected and attacked by apes that were previously on friendly terms with them. The deviant animal becomes an outsider—one of "them" rather than one of "us."[69]

Biologists use the theory of evolution to explain this pattern of hostility toward strangers. Natural selection—the engine of evolution—favors traits that help an animal's genes to proliferate. Genes are spread in two ways. One is by an animal reproducing: mating and producing offspring. The other is by an animal helping its relatives to reproduce. Because close kin share many of an individual's genes, giving blood relations a hand in the struggle for existence is an excellent way to promote one's genetic interests. That's why evolution favors kin altruism— behaviors that are geared toward promoting the well-being of an animal's relatives.

Kin altruism is a cornerstone of social behavior. Animal communities are breeding groups, and consequently fellow community members are more often than not blood relations. In light of this, loyalty to the group and hostility to outsiders makes elegant biological sense. Threatening or attacking trespassers prevents them from horning in on precious resources like food, water, and mates—resources best reserved for one's own.

It's obvious that cultural pseudospeciation resembles nonhuman xenophobia. E. O. Wilson seems to have been the first biologist to put his finger on this, commenting in his 1978 book *On Human Nature* that

> Erik Erikson has written on the proneness of people everywhere to perform pseudospeciation, the reduction of alien societies to the status of inferior species, not fully human, who can be degraded without conscience. Even the gentle San of the Kalahari call themselves !Kung—the human beings. These and other of the all-too-human predispositions make complete sense only when valuated in the coinage of genetic advantage.[70]

And just a year later, Irenäus Eibl-Eibesfeldt observed:

The formation of species . . . has its counterpart in cultural pseudospeciation. Cultures mark themselves off from each other as if they were different species. . . . To emphasize their differences from others, representatives of different groups describe themselves as human, while all others are dismissed as nonhuman or not fully equipped with all the human values. This cultural development is based on biological preadaptations, above all, on our innate rejection of strangers, which leads to the demarcation of the group.[71]

The biological explanation of xenophobia supplies a missing link in the story of dehumanization. Like many other social animals, human beings often live in mutually antagonistic communities. But dehumanization goes beyond the hatred and fear of strangers. It adds a fresh ingredient to the unit, one that is uniquely human. Thinking of others as subhumans requires sophisticated cognitive machinery. Minimally, it depends on the ability to deploy abstract concepts like "human" and "subhuman"—something that is well beyond the reach of even the cleverest nonhuman primates. More generally, dehumanization is bound up with the intricacies of symbolic culture, including notions of value, hierarchy, race, and the cosmic order. It is something that only a human brain could concoct. Although chimpanzees don't have the mental horsepower to consider their neighbors as *Unterchimpen*, there are indications of a primitive precursor of dehumanization—let's call it "despeciation"—in their behavior. We can never know for certain whether chimpanzees ever view their conspecifics as less than chimps. After all, they can't tell us. But it's possible to make some tentative inferences. Animals behave very differently when interacting with conspecifics than they do when stalking prey, even when the social interactions are aggressive or violent. Same-species aggression typically involves a lot of posturing. It enacts a choreography designed to intimidate the opponent, often by making loud, threatening sounds or puffing themselves up so as to appear as formidable as possible. The hunter's dance is completely different. A predator tries to be undetectable. It is silent and stealthy, flattening its body to the ground rather than enlarging it, and creeping forward slowly and carefully before launching itself into the final deadly sprint.

Chimpanzees are hunters. Their favorite prey is the red colobus monkey, whose flesh they consume with relish. Chimps also conduct violent raids against neighboring chimpanzee communities. To do this, they form small bands that enter another troop's territory and kill any individuals that they find and are able to overpower. They "hunt" for other chimpanzees in much the same way that they hunt for colobus monkeys.[72]

Observations like these led Jane Goodall to speculate about the relationship between pseudospeciation, dehumanization, and chimpanzee violence in her fascinating book *Through a Window: My Thirty Years with the Chimpanzees of Gombe*.

> Among humans, members of one group may see themselves as quite distinct from members of another, and may then treat group and non-group individuals differently. Indeed, non-group members may even be "dehumanized" and regarded almost as creatures of a different species. . . . Chimpanzees also show differential behavior toward group and non-group members. . . . Moreover, some patterns of attack directed against non-group individuals have never been seen during fights between members of the same community— the twisting of limbs, the tearing off of strips of skin, the drinking of blood. *The victims have thus been, to all intents and purposes, "dechimpized," since these patterns are usually seen when a chimpanzee is trying to kill an adult prey animal—an animal of another species.*[73]

Eibl-Eibesfeldt also recognized a connection between dehumanization and war, in primitive cultures as well as the developed world. His approach was subtly different from that of other pseudospeciation theorists. Whereas Erikson, Lorenz, Wilson, and Goodall all describe dehumanization as a feature of pseudospeciation, which is itself seen as a consequence of our natural tendency toward ethnocentrism and xenophobia, Eibl-Eibesfeldt suggests that it has a special role to play in war. For war to take place, he notes, human beings need to find ways to overcome biological inhibitions against lethal aggression. Dehumanizing the enemy is a means for doing this.

> In tribal societies as well as western civilization this is done through attempts to "dehumanize" the enemy. In addition, in technically advanced societies, it is done by creating deadly weapons that act quickly and at a distance. In both cases, indoctrination transfers the aggressive act to a context of being directed against another species. The opponents are degraded to inferior beings. War is primarily a cultural institution, even though it utilizes some innate dispositions.[74]

This is a subtle, multilayered analysis. We are innately biased against outsiders. This bias is seized upon and manipulated by indoctrination and propaganda to motivate men and women to slaughter one another. This is done by inducing men to regard their enemies as subhuman creatures, which overrides their natural, biological inhibitions against killing. So dehumanization has the specific function of unleashing aggression in war. This is a cultural process, not a biological one, but it has to ride piggyback on biological adaptations in order to be effective.

We will probably never know what goes on inside a chimpanzee's mind when it kills a member of its own species (although we will revisit the issue in Chapter Seven). We are better placed for finding out what occurs in human minds in similar circumstances. In the next chapter, we will explore some more recent contributions to the psychology of dehumanization, and use these to bring the dehumanizing mind into sharper focus.

3

CALIBAN'S CHILDREN

On the other side of the ocean there was a race of less-than-humans.

—JEAN-PAUL SARTRE, PREFACE TO *THE WRETCHED OF THE EARTH,* BY FRANZ FANON[1]

WE HAVE SEEN THAT EUROPEAN EXPANSION, and the intercourse with alien cultures that this entailed, encouraged philosophers and scientists to think about dehumanization. In this chapter we will look at the role that dehumanization played in the European conquest of the New World. The interlopers needed to find a place in their conceptual topography for the indigenous peoples whom they encountered. They needed to determine their position on the great chain of being. Were they fully human, subhuman, or something in between?[2]

In Shakespeare's magnificent allegory of colonialism, *The Tempest,* the character of Caliban personifies the liminal status of Native Americans. A ship's crew is marooned on an island somewhere in what Shakespeare calls the "brave new world," where they find and enslave Caliban, an entity who is presented as barely human—a "howling monster," "abominable monster," "man-monster," "a thing most brutish," "filth," a "thing of darkness," "not honour'd with human shape." Subhumanity is *intrinsic* to Caliban, for he is "a devil, a pure devil, on whose nature nurture can never stick." But there are also strains of humanity in Caliban, who suffers from his mistreatment, is aware that he is exploited, and who recognizes that he has been robbed of birthright. After four centuries, Caliban remains an iconic representation of the colonized. "Our symbol," declared Cuban poet Fernández Retamar, on behalf of present-day *mestizos*, ". . . is . . . Caliban. . . . I know of no other metaphor more expressive of our cultural situation, of our reality. . . . [W]hat is our history, what is our culture, if not the history and culture of Caliban?"[3]

After discussing the conquest and colonization of North and South America, I will use this as a springboard for extending the analysis of dehumanization. By the end, I will have uncovered an essential feature of dehumanization, one that I will build upon in the chapters to follow.

DEATH IN A BRAVE NEW WORLD

Malignant lividities and putrid ulcers often grow in the human soul, that no beast becomes at the end more wicked and cruel than man.

—POLYBIUS, *THE HISTORIES*[4]

The story of the American holocaust begins, like so many stories in the history of Europe, with the Jews. "After having expelled the Jews from your realms and dominions . . . ," wrote Christopher Columbus at the beginning of his first voyage, "your Highnesses ordered me to proceed with a sufficient fleet to the said regions of India."[5] Columbus addressed these words to King Ferdinand and Queen Isabella of Spain who had, just over a month earlier, confronted the Jews of Spain with a gut-wrenching dilemma, ordering them to convert to Christianity or leave the country. Those who chose to leave rather than betray their faith would have all of their valuables confiscated, and anyone who remained but did not comply with the edict would be put to death. Most decided to leave. They knew that as nominal Christians, they would continue to live in terror. Constantly under suspicion of practicing their religion clandestinely, many *conversos* had already been tortured and executed. The Christians despised them, and called them *Marranos*—pigs. Jews were not the only targets. Christian armies had only recently won back southern Spain from the Moors and, like the *Marranos*, the *Moriscos*—Muslims who ostensibly converted to Christianity—were enshrouded in an aura of distrust and contempt. Referred to as wolves, ravens, dogs, and evil weeds, a little over a century later they, too, would become victims of wholesale ethnic cleansing.[6]

When Columbus set sail from Palos de la Frontera in southwestern Spain, the harbor was glutted with Jewish refugees trying to escape before the final curtain fell. "It was pitiful to see their sufferings," wrote one eyewitness. "Many were consumed by hunger. . . . Half-dead mothers held dying children in their arms."[7] Columbus turned his back to the human catastrophe unfolding in Europe, and headed west. But he was unknowingly bound for a vastly more hideous calamity; one that he himself was destined to foment.

In October, Columbus made landfall somewhere in the Bahamas, and then sailed on to plant the Spanish flag on an island that he named "the Spanish land" (Hispaniola). On his triumphant return to Spain, along with half a dozen captive natives to exhibit to his countrymen, Columbus was awarded the title of Admiral of the Ocean Sea and appointed viceroy and governor of all the new lands that he had discovered. He became an instant celebrity.

Columbus set out again a year later with a fleet of seventeen ships, more than 1,200 men, and a pack of twenty dogs. This time, deadly microbes came along for the ride. Disease ravaged the Caribbean islands, as it would later decimate the rest of the Americas, and untold numbers of Indians died. The Spaniards also brought carnage. Equipped with the finest instruments of death that fifteenth-century technology had to offer, they killed, raped, and pillaged the islands. Even their greyhounds and mastiffs were trained to attack and disembowel Indians on command. As a reward, they were permitted to gorge on the flesh of their human prey.[8]

Decades later, a Dominican missionary named Bartolomé de Las Casas chronicled the Spanish depredations. "Once the Indians were in the woods," he wrote in a typical passage from his monumental *History of the Indies,* "the next step was to form squadrons and pursue them, and whenever the Spaniards found them, they pitilessly slaughtered everyone like sheep in a corral."

> It was a general rule among Spaniards to be cruel; not just cruel, but extraordinarily cruel so that harsh and bitter treatment would prevent Indians from daring to think of themselves as human beings. . . . So they would cut an Indian's hands and leave them dangling by a shred of skin and they would send them on saying "Go now, spread the news to your chiefs." They would test their swords and their manly strength on captured Indians and place bets on the slicing off of heads or the cutting of bodies in half with one blow.[9]

Las Casas had firsthand knowledge of the colonial project. Born in 1484, as a nine-year-old boy he was one of the throng that lined the streets to watch Columbus parading captive Indians through Seville. A year later, his father Pedro sailed with Columbus on his second voyage, and returned with an Indian slave as a gift to his son.[10] According to one eyewitness "we gathered together in our settlement 1,600 people male and female of those Indians . . . of whom . . . we embarked . . . 550 souls. Of the rest who were left the announcement went around that whoever wanted them could take as many as he pleased." Around 600 of the remaining Indians were taken as slaves. Most of the original 550 captive Indians died en route, and were thrown overboard for the sharks to eat. Historian David Stannard remarks, "No one knows what happened to those six hundred or so left-over natives who were enslaved, on the Admiral's orders, by 'whoever wanted them'. . . ." However, we do know that one of them reached Spain alive, and was given to the adolescent Bartolomé de Las Casas as a personal servant.

In 1502, father and son traveled together to Hispaniola, where Pedro had been awarded an *encomienda*—a grant of land and a supply of Indian slaves—for his service to the Crown. As a young man, Las Casas witnessed Indians being worked to death in mines and on plantations. Later, he would describe how newborn babies died from malnutrition, and how women would kill their infants, or induce miscarriage, to spare them such suffering.[11]

Las Casas also witnessed the atrocities of war. As a chaplain during the invasion of Cuba, he looked on as:

> A Spaniard . . . suddenly drew his sword. Then the whole hundred drew theirs and began to rip open the bellies, to cut and kill those lambs—men, women, children, and old folk, all of whom were seated, off guard, and frightened. . . . The Spaniards enter the large house nearby, for this was happening at its door, and in the same way, with cuts and stabs, begin to kill as many as they found there, so that a stream of blood was running, as if a great number of cows had perished.[12]

When Columbus made landfall in 1492, Hispaniola was home to around a million people. By 1510, only eighteen years later, a lethal cocktail of virgin soil infection and Spanish oppression had reduced them to 46,000. In 1509, something between 600,000 and one million souls inhabited the neighboring islands of Jamaica, Cuba, and Puerto Rico, but by 1552 no more than two hundred were left.[13] At the same time, similar events were unfolding all over the New World—in Mexico, Venezuela, Brazil, Peru, Florida, and elsewhere. One eyewitness reported that in Panama the Spaniards hacked off limbs "like butchers cutting up beef and mutton for market," and another complained that "some Indians they burned alive; they cut off the hands, noses, tongues, and other members of some; they threw others to the dogs; they cut off the breasts of women."[14] A group of Dominican friars protested to the future king and Holy Roman Emperor Charles V that:

> Some Christians encounter an Indian woman, who was carrying in her arms a child at suck; and since the dog they had with them was hungry, they tore the child from the mother's arms and flung it still living to the dog, who proceeded to devour it before the mother's eyes . . . when there

were among the prisoners some women who had recently given birth, if the new-born babes happened to cry, they seized them by the legs and hurled them against the rocks, or flung them into the jungle so that they would be certain to die there.[15]

In 1511, Las Casas heard a priest named Antonio Montesinos preach against the Spanish barbarities. "I am the voice crying in the wilderness . . . ," Montesinos told his congregation, "the voice of Christ in the desert of this island . . . [saying that] you are all in mortal sin . . . on account of the cruelty and tyranny with which you use these innocent people. Are these not men? Have they not rational souls? Must not you love them as you love yourselves?"[16] After listening to Montesinos, Las Casas experienced a crisis of conscience, gave up his slaves, and campaigned for Native American rights—tirelessly writing, preaching, and petitioning the Crown on their behalf. Because of his efforts, the King of Spain, probably the most powerful man in the world at the time, ordered that a debate take place between Las Casas and a man named Juan Ginés de Sepúlveda on the legitimacy of using force against the Indians. It took place in the northern Spanish city of Valladolid in the year 1550.

Sepúlveda was a well-known humanist and Aristotelian scholar, and he based his case on Aristotle's theory of natural slavery, arguing that Native Americans are "slaves by nature, uncivilized, barbarian and inhuman." In approaching the matter in this way, Sepúlveda was part of a scholarly tradition that had been going on since at least 1510, when the Scottish philosopher and theologian John Mair described the indigenous people of the Caribbean as natural slaves who "live like beasts" (Mair's work was well known in Spain, and was cited in Spanish debates about Native American rights). However, Sepúlveda pressed the idea of Indian barbarism further than his predecessors had done. He insisted that there is almost as great a difference between Indians and Spaniards as between monkeys and men, and assured the jury that "you will scarcely find even vestiges of humanity" in them, and that, although the natives are not "monkeys and bears," their mental abilities are like those of "bees and spiders." Why bees and spiders? This is probably a reference to a passage from Aristotle's *Physics*. The context is a discussion of purposive behavior by animals. Behavior that is purposive but nonetheless irrational is evident "in the case of animals other than man, since they use neither craft nor inquiry nor deliberation in producing things—indeed this is why some people are puzzled about whether spiders, ants, and other such things operate by understanding or in some other way." Aristotle believed that only humans can think. So, in comparing the behavior of Indians to that of spiders and ants, Sepúlveda implicitly denied that they are rational—and therefore human—beings.[17]

Sepúlveda also referred to the Indians as *homunculi*. The notion of the homunculus was a fixture of the medieval imagination. Homunculi were thought to be humanoid entities produced in an unnatural manner from human sperm. There were two theories of how homunculi come into being. Some alchemists claimed to be able to create homunculi in the laboratory, rather like medieval test-tube babies. A writer known as Pseudo-Thomas (thus named, because he tried to pass his writings off as authored by Thomas Aquinas) described an experiment in which one "takes the semen of a man and places it in a clean vessel under the heat of dung for thirty days" after which "a man having all the members of a man is generated there." Pseudo-Thomas claimed that, although this creature would outwardly resemble a human being, it would not possess a human soul (Pseudo-Thomas lifted this story from the writings of the ninth-century Arab alchemist Abū Bakr Muhammad Ibn Zakarīyā al-Rāzī). Writing in a similar vein, the Spanish theologian Alonso Tostado described an experiment supposedly performed by the Spanish alchemist/physician Arnald of Villanova, who sealed some human semen in a container with some unspecified drugs. He reported, "Finally after some days, many transmutations having occurred, a human body was formed out of it, but not perfectly organized." Arnald reportedly destroyed his homunculus by smashing the vessel in which it was growing, because he was uncertain whether God would infuse a human soul into the artificial being that he had created.[18]

At the time of the debate in Valladolid, the world authority on homunculi was none other than Paracelsus, the grandiose Swiss physician and alchemist whom we met in Chapter Two. Although he doubted that homunculi could be produced in the lab, Paracelsus accepted their reality. He expounded on the topic in a tract entitled *De homunculis*, written circa 1530. You may recall that Paracelsus believed that human beings possess both an animal and a spiritual nature. Homunculi are offspring of the animal component. As William R. Newman explains it, in his fascinating book *Promethean Ambitions: Alchemy and the Quest to Perfect Nature*, "the animal body of man exists independent of the soul, and it produces a defective, soulless sperm when one is possessed by it. Paracelsus . . . tells us, that homunculi and monsters are produced: *therefore they have no soul.*"[19] Homunculi are generated from decaying semen. When a man experiences lust, animal sperm is produced, which either "putrefies" inside of him if not ejaculated, resulting in the growth of homunculi internally, or—if he discharges his sperm through lustful intercourse—causes homunculi to grow in the body of his sexual partner (oral sex causes homunculi to grow in the throat, and anal sex produces them in the intestines).

Paracelsus proposed a radical intervention for men anxious to avoid producing soulless offspring.

> [I]f a man wants to keep himself chaste by force, and relying on his own strength, he should be castrated or castrate himself, that is, dig out the fountain where that lies of which I write. Therefore God has formed it . . . in front of the body on the outside.[20]

We can't be certain why Sepúlveda called Native Americans homunculi, but it seems likely that he was trying to convey the idea that they did not have human souls. Sepúlveda's image of the Indians was not exceptional. Fourteen years earlier, another missionary and advocate of Indian rights, Bernardino de Minaya, complained that the Spanish considered Indians as "not true men, but a third species of animal between man and monkey created by God for the better service of man," while others denigrated them as "talking animals" and "beasts in human form." Sepúlveda simply put an academic gloss on preexisting bigotry, as did other scholars of the day who proposed that Indians were not descended from Adam and Eve, but were formed from the decaying debris left behind by the Great Flood. Those who denied the Indians' humanity on the grounds of their non-Adamic lineage included the philosopher Giordano Bruno, the physicians Andrea Cesalpino and Paracelsus, and the mathematician Gerolamo Cardano.[21]

Las Casas didn't dispute the Aristotelian theory that barbarians are natural slaves, but he challenged the claim that Native Americans were barbarians, addressing Sepúlveda's arguments point by point. The debate continued for about a month. During this time, the two men never confronted one another in the flesh. Each separately presented his case to a fourteen-man jury that was appointed by the king. Although there's no record of their decision, Sepúlveda later wrote that all but one of them supported his position.

Dramatic though it was, the debate at Valladolid was inconclusive, and had no discernible impact on Spanish colonial policy. However, Las Casas's relentless campaign for Indian rights, and his arguments that they were human beings, bore fruit. Thanks to his efforts, Pope Paul III proclaimed in 1537 that the Indians were human beings with rational souls and therefore should not be enslaved, and Las Casas was also instrumental in bringing about sweeping reforms of the ecomienda system of quasi-slavery, which, in principle if not in practice, gave indigenous people some protection from oppression and abuse.[22]

VIRGINIA, MASSACHUSETTS, AND BEYOND

If dogs were trained up to hunt Indians as they do bears, we should be quickly sensible of a great advantage thereby. . . . The dogs would do a great deal of execution upon the enemy and catch many an Indian that would be too light of foot for us.

—REV. SOLOMON STODDARD, LETTER
TO GOVERNOR JOSEPH DUDLEY[23]

In North America, settlers put down roots along the northeast coast of what is now the United States just over a century after the Spanish established their first permanent settlement in the Caribbean.

The story begins in 1607, when an enterprising group of English businessmen called the London Company established a colony in what is now Virginia. They called it Jamestown.

At first, the settlers failed to thrive. Food shortages led to famine so severe that they turned to cannibalism. Two-thirds of them died during the first year, but by 1619 the colony had become the center of a burgeoning and lucrative tobacco industry. A steady stream of vessels from England disgorged their cargo of would-be entrepreneurs looking to get a cut of the action, and indentured servants hoping to create a new life in a new world. As the population swelled, plantation owners cleared and cultivated land at a feverish pace in their efforts to satisfy the new European craving for tobacco, and as their footprint grew larger, tensions between the settlers and Native Americans escalated, reaching a boiling point in 1622, when Powhatan warriors attacked settlements along the James River, killing about a quarter of the inhabitants, including elderly men, women, and children.

The massacre of 1622 marked a turning point in English-Indian relations. Previously, aggression against the Indians had been sporadic, but after 1622 it became policy. It was at this point that the dehumanization of Native Americans began to get real traction. Captain John Smith set the tone in his description of the Indians as "cruell beasts" with "a more unnatural brutishness than beasts," while back in England, Samuel Purchas, a well-known compiler of travel books, informed his readers that Indians are organisms "having little of Humanitie but shape . . . more brutish than the beasts they hunt, more wild and unmanly than that unmanned wild Country, which they range rather than inhabite." The poet Christopher Brooke was even more explicit, describing the Indians of Virginia as "creatures," and adding, by way of explanation, "I cannot call them men." He made a point of casting them as "of inhuman birth," "dregs," and "garbage," and asserted that they were not of the lineage of Adam and Eve, but had "Sprung up like vermine of an earthly slime." The colonists also enlisted Aristotle in their campaign, calling Indians "barbarous" and "naturally born slaves."[24]

Meanwhile, English Puritans were busy carving out a new life in Massachusetts. Brimming with religious enthusiasm, they initially targeted the Indians as heathens ripe for conversion to Christianity, but as the white population increased and competition for resources intensified, missionary zeal gave way to genocidal hostility. The first all-out war between American colonists and Native Americans (King Philip's War) erupted when settlers began moving into Pequot Indian territory in Connecticut. In its culminating episode, Englishmen joined Narragansett and Mohegan Indians, surrounding a Pequot village near Mystic, Connecticut, and setting it on fire. All of the inhabitants—some 800 to 900 men, women, and children—were killed, many of them burned alive. Not long after the English victory, when most of the remaining Pequot had been mopped up and either executed or sold into slavery in the Caribbean, Plymouth governor William Bradford wrote exultantly that the burning bodies and streams of blood

> seemed a sweet sacrifice, and they gave the praise thereof to God, who had wrought so wonderfully for them, thus to enclose their enemies in their hands and give them so speedy a victory over so proud and insulting an enemy.[25]

Once again, the dehumanizing process took its deadly course. Wait Winthrop openly expressed the new sensibility in a 1675 poem celebrating the hoped-for extermination of Native Americans. Still reeling from the defeat of colonial troops by Narragansett Indians in the battle of Great Swamp, Winthrop encouraged his readers to look forward to the happy day when the Indians, who were variously described as "flies," "rats," "mice," and "swarms of lice," would be driven to extinction.[26] Winthrop's aspirations were widely shared for centuries to come. Around the time of the American Revolution, the Indians (who sided with the British) were castigated as "copper Colour'd Vermine" fit to be "massacre[d] to such a degree that [there] may'nt be a pair of them left, to continue the Breed upon the Earth" and a decade later a British visitor to the newly minted republic reported that white Americans "have the most rancorous antipathy to the whole race of Indians; and nothing is more common than to hear them talk of extirpating them totally from the face of the earth, men, women, and children."[27]

Native Americans were alternately—and sometimes simultaneously—depicted as vermin, human refuse, and as deadly nonhuman predators. The English king James I, after whom Jamestown was named, called them "beastly . . . slaves to the Spaniards, refuse of the world." "Once you have but got the track of those Ravenous howling Wolves," advised Cotton Mather, "then pursue them vigorously; turn not back till they are consumed."[28]

Whereas at the inception of their American adventure, the Puritans had considered the Indians to be degenerate human beings snared in the devil's clutches, it didn't take long for them to cast the Indians as devils incarnate. The "red devils" of this ethnic demonology were said to possess telltale predatory traits—they were "untamed," "cruel," and "bloodthirsty" (the "merciless Indian savages" of the Declaration of Independence). In rhetoric reminiscent of Cotton Mather, George Washington informed a correspondent that Indians and wolves are both "beasts of prey, tho' they differ in shape."[29]

By the dawn of the nineteenth century, the image of the Indian as predator had gained broad currency. As John Wakefield, a U.S. Army private who chronicled the 1832 Black Hawk War, succinctly put it, Indians are "most like the wild beasts than man." In the same vein, Pennsylvania Supreme Court Judge Hugh Henry Brackenridge referred to "animals, vulgarly called Indians," and Francis Parkman, one of greatest American historians of the nineteenth century, unblushingly described in his 1847 book *The Oregon Trail: Sketches of Prairie and Rocky Mountain Life* how dehumanizing the Indians made their extermination morally acceptable.

> For the most part a civilized white man can discover very few points of sympathy between his own nature and that of an Indian. . . . Nay, so alien to himself do they appear that . . . he begins to look upon them as a troublesome and dangerous species of wild beast, and if expedient, he could shoot them with as little compunction as they themselves

would experience after performing the same office upon him.[30]

By the midcentury, American settlers in Arizona had embarked on what was explicitly described as a war of extermination against the "Apache" (a name that whites used indiscriminately for all of the Indians of the region). Settlers did not merely regard these people as animals. They were characterized as superlatively dangerous predators—"the most savage wild beast"—and were avidly hunted. "Persons were constantly coming in," observed Judge Joseph Pratt Allen, "who wished to join the party, one and all believing and talking of nothing but killing Indians. . . ." In accord with their subhuman status, the corpses of dead Indians were treated like game. On one hunting expedition, one participant recorded that the brains of five dead Apache were used to treat a deer hide to make buckskin ("the best buckskin I ever seed was tanned with Injun brains," he remarked). Both civilian Indian hunters and U.S. military personnel most often characterized the Apache as wolves, just as George Washington and others had done a century before.[31]

> Expeditions became in many military dispatches "hunts"; the Apache inevitably "wolves." The 1867 report of the U.S. Secretary of War, for example, referred to fighting Apaches as "more like hunting wild animals than any regular kind of warfare" and noted that the Apaches "like wolves . . . are ever wandering." As the U.S. Army officer Davis Britton, posted to Arizona a decade later put it, "[W]e hunted and killed them as we hunted and killed wolves."[32]

DEFINING THE HUMAN

The history of the relationship between European settlers and Native Americans illustrates how dangerous dehumanization can be. The acts of violence perpetrated by settlers, as well as their tendency to turn a blind eye to the suffering that they caused, were intertwined with their conception of Indians as less than human.

Today, every educated person knows that we are all *Homo sapiens*—members of a single species—and that the biological differences between one human group and another are trivial at most. But the dehumanizing impulse operates at the gut level, and easily overrides merely intellectual convictions. To understand why it has this power, we've got to answer a fundamental question. What *exactly* goes on when we dehumanize others?

The obvious place to look for an answer is in the writings of psychologists. If anybody understands the psychology of dehumanization, it should be them. However, as Australian psychologist Nick Haslam has recently pointed out, researchers have for the most part neglected the most fundamental question. "Any understanding of dehumanization," he observes, "must proceed from a clear sense of what is being denied to the other, namely humanness. However . . . writers on dehumanization have rarely offered one."[33]

Can't we turn to science for an explanation of humanness? Surely, one might think, biologists can tell us exactly what it means to be human. Actually, they can't, for a reason that is often overlooked. *Human* doesn't have any fixed meaning in biology. Some scientists equate *human* with both modern human beings and Neanderthals, while others speak about the split between Neanderthals and humans (in which case humans are equated with *Homo sapiens sapiens*). Others describe all members of the genus *Homo* as human, while still others reserve the term *human* for all of the species in our lineage after our common ancestor with the chimpanzee.[34] In short, biologists' use of *human* is all over the map. The reason for this is a simple one. *Human* belongs to a completely different taxonomy—a pre-Darwinian folk-taxonomy that owes more to the great chain of being than it does to modern biological systematics. The two frameworks are incommensurable.

To appreciate the depth of the problem, let's do a little thought experiment. Suppose that on a planet that is very much like Earth (call it "Schmearth") orbiting a distant star, creatures evolved that were anatomically, physiologically, and behaviorally indistinguishable from

Homo sapiens. In fact, we can imagine that each earthling alive today has a schmearthling counterpart. However, earthlings have a completely different evolutionary history from schmearthlings. Life emerged independently on the two planets, but once it emerged, evolution took the same course, stumbling on the same anatomical structures and physiological mechanisms (what philosopher Daniel C. Dennett calls "good tricks").[35] As a result, if you were teleported to Schmearth, à la *Star Trek*, it would seem just like Earth.

In this scenario, no earthling would have an ancestor in common with any schmearthling. Earthlings and schmearthlings would be completely unrelated to one another, even though there would be no way to tell them apart.

What, exactly, is the relationship between earthlings and schmearthlings? If earthlings are human and schmearthlings are indistinguishable from earthlings, then it seems like we ought to say that schmearthlings are human. It would seem churlish to deny humanity to someone exactly like you just because of where they happened to be born. Are schmearthlings *Homo sapiens*? No, they're not. Any biologist will tell you that every member of the same species must be part of the same biological lineage—descended from the same ancestor. So, it seems reasonable to say that if schmearthlings existed, they would be humans but not *Homo sapiens*. And if this is right, then it's not true to say that all humans are *Homo sapiens*.

Harvard psychologist Herbert C. Kelman was one of the first, if not the first, psychologists to use the term *dehumanization* in a scientific context. He pointed out in a landmark article in 1973 that to understand dehumanization, we need to know what it is to perceive another person as human."[36] Kelman wisely refrained from speculating about what it is to be human. Instead, he wanted to illuminate the folk-concept of the human. "I would propose," he wrote, "that to perceive another as human we must accord him identity and community. . . ."

> To accord a person identity is to perceive him as an individual, independent and distinguishable from others, capable of making choices, and entitled to live his own life accord-

ing to his own goals. To accord a person community is to perceive him—along with oneself—as part of an interconnected network of individuals, who care for each other, who recognize each other's individuality, and who respect each other's rights. These two features together constitute the basis for individual worth.[37]

Kelman went on to explain some of the implications of his analysis, writing that:

To perceive others as fully human means to be saddened by the death of every single person, regardless of the population, group, or part of the world from which he comes, and regardless of our own personal acquaintance with him. If we accord him identity, then we must individualize his death [and] . . . if we accord him community, then we must experience his death as a personal loss.[38]

As UCLA sociologist Leo Kuper remarks, Kelman offers an extremely idealized notion of what it is to regard another person as "fully human."[39] Many people die every day, often under horrendous circumstances, but very few of us lose any sleep over this. We certainly don't feel these deaths as a personal loss, yet we don't consider the victims as subhuman either. However, when shorn of its excesses, Kelman's definition seems to point in the right direction.

Consider his notion of community. It's true that dehumanized others are socially marginalized—the dehumanized person is not one of us (whoever "us" happens to be). Characteristics that set them apart from the majority are emphasized, a task that is most easily accomplished if obvious differences like skin color come into play. As I mentioned in Chapter One, predominantly white Allied troops treated Japanese people as subhuman much more extensively than their German counterparts. Likewise, Native Americans, African Americans, and Chinese immigrants were all easy targets because their physical characteristics made them stand out from the white majority.

Sometimes the differences aren't quite so obvious. In these cases, social practices are implemented to isolate the dehumanized group. The historical relationship between Muslims and Jews in North Africa and the Middle East provides an interesting case. Thriving Jewish communities existed in Muslim lands for many centuries. Both Jews and Muslims were of more or less identical stock—they were Semites. So, social rituals and symbols were needed to differentiate Jews from their Muslim overlords.

Jews were accorded the status of *dhimmi*—people permitted to dwell in the House of Islam in a condition of subservience. The Koran stipulates that *dhimmi* are to be granted basic liberties, including freedom of worship, but they must pay a special tax (*jizya*) and live in "willing submission" to the Muslim majority.[40] Although initially undefined, this "willing submission" eventually came to include a range of social restrictions and humiliations, some of which were itemized by the eleventh-century Persian theologian Abu Hamid Muhammad ibn Muhammad al-Ghazali as follows:

> The *dhimmi* must . . . pay the *jizya* . . . on offering up the *jizya*, the *dhimmi* must hang his head while the official takes hold of his beard and hits [the *dhimmi*] on the protuberant bone beneath his ear. . . . [T]heir houses may not be higher than the Muslim's, no matter how low that is. The *dhimmi* may not ride an elegant horse or mule; he may ride a donkey only if the saddle is of wood. He may not walk on the good part of the road. They [the *dhimmis*] have to wear an [identifying] patch [on their clothing].[41]

Within a few centuries, the practice of requiring Jews to wear a special uniform was adopted all over Christian Europe (it was declared mandatory by Pope Innocent III in 1215). Jews were required to wear a conical hat and a yellow Star of David—yellow to symbolize Judas Iscariot's betrayal of Jesus for gold (oddly, in light of the fact that Judas was supposed to have betrayed Jesus for thirty pieces of silver).[42] After their invasion of Poland in 1939, the Nazis revived this practice, compelling

Jews to sew a yellow Star of David inscribed with the word "Jude" ("Jew") on their clothing. Hitler's Germany provides an exceptionally clear illustration of the ritualistic paraphernalia of social exclusion. As Duke University historian Claudia Koonz points out in her book *The Nazi Conscience*:

> Nazism offered all ethnic Germans . . . a comprehensive system of meaning that was transmitted through powerful symbols and renewed in communal celebrations. It told them how to differentiate between friend and enemy, true believer and heretic, Jew and non-Jew. In offering the faithful a sanctified life in the Volk, it resembled a religion. Its condemnation of egoism and celebration of self-denial had much in common with ethical postulates elsewhere. But in contrast to the optimistic language of international covenants guaranteeing universal rights to all people, Nazi public culture was constructed on the mantra "Not every being with a human face is human."[43]

Kelman's point that dehumanized people are shorn of their individuality also rings true. They are typically thought of as fungible, as parts of an undifferentiated mass. Propagandists often exploit this frightening image. Perhaps the most notorious example is a scene from the notorious German film *The Eternal Jew* portraying Jews as a seething swarm of rats. Similar conceits were used in propaganda cranked out by other nations, including the United States. A 1945 film called *Japan: Know Your Enemy*, directed by Frank Capra (who directed several popular motion pictures during the 1930s and '40s, including *Mr. Deeds Goes to Town*, *Mr. Smith Goes to Washington*, and *It's a Wonderful Life*), presented Japanese people as indistinguishable from one another, "photographic reprints off the same negative." George Orwell expressed the position even more explicitly. When in Marrakech, he wrote, "it is always difficult to believe that you are walking among human beings . . . Are they really the same flesh as yourself? Do they even have names? Or are they merely a kind of undifferentiated brown stuff, about as individual as bees or coral insects?" The disconcertingly fecal image of Moroccans as "undifferentiated brown stuff" has a counterpart in imagery used more recently in discussions of illegal immigration from Latin America to the United States, a country alleged to be "awash under a brown tide" of Mexican immigrants (as almost a century earlier, the American anti-immigrationist Lothrop Stoddard had warned that white America was soon to be swamped by a "rising tide of color"). The significance of the expression "brown tide" may not be obvious to all readers. The term refers to an algae infestation specific to the Gulf of Mexico that turns seawater brown.[44]

More recently still, Norberto Ceresole, an adviser to Venezuelan president Hugo Chavez, expressed the principle with chilling frankness. Ceresole recounts his epiphany that Jews masterminded the 1994 bombing of the Argentine Jewish Mutual Aid Association in Buenos Aires, which killed eighty-five people and injured three hundred. The Jews were, he said, "not as I had known them until then, that is as individuals distinct from one another, *but rather as elements for whom individuation is impossible* . . ." At the time of writing there have been no convictions for this crime, although Hezbollah operatives are suspected.[45]

Research findings support anecdotal examples like the ones I have just described. Social psychologists confirm that we are likely to perceive people outside our own community as more alike than those within it. We perceive members of our own group as individuals, but see other groups as more or less homogenous (psychologists call this the "outgroup homogeneity bias"). When the outgroup homogeneity bias merges with the outgroup bias described in Chapter Two, the result is a dangerous mix. Outsiders are both denigrated and stereotyped: we are a richly diverse community of praiseworthy individuals, but they are all dishonest, violent, filthy, stupid, or fanatical.[46]

Kelman's analysis of the concept of the human has a lot in its favor, but it also has some major shortcomings. Before considering them, it will be useful to reflect on what's required for the satisfactory analysis of a concept. When developing an analysis of a concept, the analysis should be such that it includes everything that ought to come under the concept, while excluding everything that is external to it. Philosophers explain this by citing what they call its necessary and sufficient conditions. Necessary conditions set out what has to be true of a thing for it to come under the concept. Suppose that you wanted to analyze the concept "porcupine." You might begin by making a statement of the form, "All porcupines are so-and-sos," for instance: "All porcupines are animals." This wouldn't be incorrect, but it wouldn't be very informative, either, because it includes too much. Sure, porcupines are animals—but so are blue whales, beagles, and butterflies. A satisfactory analysis has got to rule out these other creatures. This is where sufficient conditions come in. Sufficient conditions set out what characteristics of a thing are enough for it to come under the concept. To specify the sufficient conditions for being a porcupine, you need a statement of the form "All so-and-sos are porcupines," for instance: "All spiny mammals native to North America are porcupines." This would certainly be true—porcupines are the only spiny mammals native to North America—but it would be overly restrictive, because it excludes those species of porcupine that are native to other parts of the world.

A really good analysis specifies conditions that are both necessary and sufficient, and thereby pinpoints exactly what comes under the concept in question. Necessary and sufficient conditions are the Holy Grail of definitions; often sought but seldom found. Mathematical and logical concepts are about the only ones that are precise enough to permit this sort of analysis. Most of our ordinary, workaday notions (as well as most scientific ones) are far too fuzzy. To appreciate the difficulties involved, try to work out necessary and sufficient conditions for the concept of *the beautiful*. What exactly is it for something to be beautiful? Sunsets, faces, music, and even equations can be described as "beautiful"—but what do all such examples (and only such examples) have in common? We might try something like "pleasing to the senses," but this doesn't work. The condition is not even a necessary one, and is miles away from being sufficient. Ideas can be beautiful, but they can't be perceived with our sense organs. And can't something be beautiful without anyone ever seeing it? Orchids were every bit as beautiful fifty million years ago as they are today, even though there was nobody around then to marvel at their splendor. Chocolate is pleasing to smell and taste, but I would hesitate to call it beautiful.

Although searching for the necessary and sufficient conditions that define everyday concepts is ultimately quixotic, it's still a worthwhile ideal, because the closer we can come to them the more precise and nuanced our understanding will become.

Bearing these points in mind, let's see how Kelman's analysis of dehumanization stacks up. The first thing to notice is that the two psychological biases we have been considering—the out-group bias and the out-group homogeneity bias—aren't specific to dehumanization. Although they play a role in the dehumanizing process, they play a role in other derogatory attitudes as well. A person might believe that the members of some ethnic group share certain undesirable characteristics—for instance, that all Arabs are violent religious fanatics. Such attitudes are regrettable, to be sure, but they fall short of *dehumanizing* Arabs. People often disparage others without denying their humanity. So, the out-group bias and the out-group homogeneity bias may be necessary for dehumanization, but they're not sufficient for it. The second thing to notice is that there are plenty of examples of dehumanization that don't involve denying that others are rational agents, as Kelman claims. Consider the Holocaust is the most thoroughly documented episode of dehumanization in history. The Nazis didn't deny that Jews were rational agents. In fact, they felt threatened by what they took to be the Jews' collective and conspiratorial agency—their destructive goals and degenerate values. Hitler's policy of extermination was based on his belief that the world was in the grip of a vastly powerful Jewish conspiracy that was implacably hostile to the spiritual and material flourishing of the Aryan race, and single-mindedly devoted to its destruction. Nothing short of mass execution could save the planet from their diabolical project of world domination.

How about the principle of fungibility? Even though the Jews were believed to be plotting against Germany, the notion that the entire Jewish race was dedicated to this project suggests that the Nazis denied that Jews were individuals (recall the swarm of rats in *The Eternal Jew*). This is perfectly true, but—once again—it's not unique to dehumanization. The very same Germans who tried to exterminate the Jews of Europe strove to subordinate their *own* individuality to the Nazi state. The ideal of the German people as an aggregate entity was clearly expressed in the 1933 inaugural address of the Nazi philosopher Martin Heidegger: "The Führer has awakened this will in the whole nation," Heidegger trumpeted, "and has fused it into one single will. . . . Heil Hitler!" The Nazis had a special term for this process of homogenization; they called it *Gleichschaltung*, which is roughly translated as "bringing into line." As one German citizen revealingly explained, *Gleichschaltung* "means that the same stream will flow through the ethnic body politic [literally, 'body of the people']."[47] But the Nazis didn't dehumanize themselves, as Kelman's reasoning would seem to imply. Instead, they considered themselves to be the purest and most exalted form of humanity.

Nick Haslam, the University of Melbourne psychologist whom I mentioned earlier, takes a somewhat different approach toward the question of what it is that dehumanized people are supposed to lack. Haslam proposes that we operate with two distinct concepts of humanness. One is that humanness consists of characteristics that only human beings possess. Take language. No other species can arrange words into sentences to communicate information, which means that the capacity for language is uniquely human (as is the ability to dance the samba or memorize the libretto of *La Bohème*). The other is that humanness consists of characteristics that are "essentially, typically or fundamentally" human. The language here is a bit confusing, as it's not quite clear what makes an attribute essentially, typically or fundamentally human (indeed, it's vacuous to define the human as "that which is essentially human"). But let's look beyond these infelicities and try to understand what it is that Haslam is gesturing toward. The idea seems to be that we have a stereotypical image of what a human being is—a kind of model or paradigm. Consider bipedalism. When we imagine a typical human being, we picture a person standing upright. In reality, of course, there are plenty of people who can't stand upright: babies, for instance, and people with certain forms of physical handicap. There are also nonhuman animals that stand on two legs (when Plato defined human beings as featherless bipeds, Diogenes of Sinope, the Harpo Marx of ancient philosophy, crashed his lecture wielding a plucked chicken and shouting, "Here is Plato's man!"). Nevertheless, there is a strong association in our minds between being human and standing on two legs.[48]

So, how are these two concepts of humanity linked with forms of dehumanization? Haslam suggests that when people are stripped of their typically human attributes they're seen as cold and inert—as inanimate objects lacking warmth, individuality, and agency. In contrast, people denuded of their uniquely human characteristics are perceived as subhuman creatures without language, incapable of reflection and refined emotions, devoid of imagination and intelligence, without culture, industriousness, or self-control.

It's certainly true that people sometimes treat others as though they were inanimate objects—as equivalent to robots, inflatable dolls, or mere statistics. But this sort of dehumanization isn't pertinent to this book. There are a couple of reasons for this. Viewing others as inanimate objects distances one from them, but it doesn't seem to motivate behavior or play a significant role in producing mass violence. Sure, the Nazi bureaucrats treated prisoners as mere numbers, but the Nazi bureaucrats treated *everyone* as numbers. That's part of the bureaucratic mind-set. The men (and occasionally women) who actually committed atrocities, as well as the leaders that commanded them to do so, emphatically did not conceive of Jews as numbers. You don't kill numbers. Israeli anthropologist Eyal Ben-Ari distinguishes "objectification" from dehumanization, which correspond to Haslam's two forms of dehumanization. Ben-Ari points out that, at least in the military context, both "us" and "them" are objectified, whereas dehumanization exaggerates the *difference* between "us" and "them." This dissimilarity suggests that the forces producing these two phenomena may be correspondingly distinct.[49]

The other form of dehumanization is much more relevant for understanding collective violence. As we've already seen, people who commit atrocities often conceive of the targets of their aggression as lower forms of life. But there's something wrong with Haslam's explanation of why this is. Think of a newborn baby. We all accept that babies are human beings. But on Haslam's analysis this is puzzling, because babies lack the uniquely human characteristics that he lists. Neonates can't speak or engage in higher order thought, their emotions are at best extremely crude, and they are not industrious, imaginative, or cultured. If we consider babies to be human even though they lack the traits dubbed "uniquely human," then it simply can't be true that anyone without these characteristics is viewed as subhuman. Pushing the point further, Shakespeare's writing of *Hamlet* was a uniquely human achievement if ever there was one. No other animal (even the proverbial multitude of monkeys pounding interminably on typewriters) could have pulled it off. But even though writing *Hamlet* was uniquely human, only William Shakespeare wrote it. It's true that anyone that wrote *Hamlet* is human, but false to say that anyone that's human wrote *Hamlet*. Writing *Hamlet* is sufficient for humanness, but it's not necessary. The same is true of the other uniquely human traits.

COUNTERFEIT HUMANS

Haslam, Kelman, and other psychologists who theorize about dehumanization make a key assumption that is probably incorrect, and which hampers their efforts. They assume that we equate the essence of a thing with its observable characteristics. Haslam, for example, lists various attributes—higher order thought, language, refined emotions, and so on—and supposes that people conceive humanness in terms of them.

Do we really think of humanness as the sum of a set of observable characteristics? Or do we think of it as something deeper that is hidden from view?

Before offering answers to these questions, I need to introduce a notion that plays a prominent role in the discussion—the concept of *natural kinds*. The concept of natural kinds (and the related notion of essences) is central to the analysis of dehumanization that I will develop later in this book. So, I will spend some time discussing it here to lay the groundwork for the more extensive discussion in Chapter Six.

Imagine that there are several dozen pieces of jewelry spread out on a table in front of you. There are rings and broaches, pendants, bracelets, and earrings. Some of these are made of silver, and some of them are made of gold. Further, imagine that you are asked to arrange these items into groups, based on the kinds of things they are. There are lots of ways that you might perform this task. One way would be to put all the earrings in one pile, all the bracelets in another, all the broaches in a third, and so on, making five piles in all. Another method would be to make just two piles, one for all the gold jewelry and the other for all the silver jewelry.

Obviously, the choice of one method or another depends on the sort classification scheme that you find most appealing. It's not that one is right and the others are wrong—the choice of a classification scheme is just a matter of how you want to cut the pie.

Suppose you use the first method. You'd be dividing the jewelry into categories based on what they're used for. Philosophers call these "artificial kinds," because they're based on human preferences and practices rather than on real divisions "out there" in nature. They are classifications of convenience. But if you performed the task in the second way—putting all the gold jewelry into one pile and all the silver jewelry into another pile—you would be classifying them as natural kinds, as categories that exist objectively in nature, independent of human artifice.

Now, let's dig a little deeper.

Suppose that one of the earrings was made from white gold, and because the earring looked like silver, you placed it in the silver pile. If you had done that, you would have made a mistake, because something's being gold or silver isn't fixed by how it looks. Objects belong in the "gold" category only if they are made from atoms with atomic number 79, and those that fall into the "silver" category must have atomic number 47 (i.e., atoms with 79 and 43 protons respectively). Atomic numbers determine what gold and silver are.

This has the startling implication that, centuries ago, before the advent of modern chemistry, nobody knew what gold and silver were! Of course, our forebears could distinguish between gold objects and silver objects, and they got it right most of the time, but they didn't know precisely what it was *about* these things that made them gold or silver. However, they had some inkling. Even in those far-off days, people suspected that there was a hidden "something" that determined whether an object was gold, silver, or some other substance. The medieval alchemists—the intellectual ancestors of today's chemists—spoke of the "souls" of metals. The soul or essence of a metal was supposed to be something distinct from its appearance or "body"—something that makes it the kind of metal that it is. Interestingly, the alchemists saw this as analogous to the soul or essence of a human being. The Polish alchemist Michael Sendivogius, for one, made this point quite explicitly in a tract entitled *The New Chemical Light*.

> The bodily nature of things is a concealing outward vesture. If you dressed a boy and a girl of twelve years of age in exactly the same way, you would be puzzled to tell which was the boy and which the girl, but when the clothes are removed they may easily be distinguished. In the same way, our understanding makes a shadow to the shadow of Nature, for our human nature is concealed by the body in the same way as the body by the clothes.[50]

The notion of the "soul" of gold (or any other natural kind) functioned as a placeholder for a mysterious unknown. Although Sendivogius's text dates from the early modern period, just a year after settlers founded the Jamestown colony, it expresses a view that originated in antiquity and, if anything, gained impetus from the intellectual revolution of the seventeenth and eighteenth centuries.

Later in the seventeenth century, the English philosopher John Locke wrote one of the most influential discussions of natural kinds and their essences.* Locke speculated in his monumental *An Essay on Human Understanding* that members of natural kinds have an unob-

---

*For the sake of accuracy, Locke didn't use the term *natural kind*, which was introduced by John Stuart Mill over a century later.

servable "real essence" that accounts for their observable properties. He illustrated the point with a gold ring. "The particular parcel of matter," Locke wrote, in his characteristically tortuous literary style, "which makes the ring I have on my finger, is . . . by most men, supposed to have a real essence, whereby it is gold."

For ordinary purposes, we say that something is gold because of its "peculiar color, weight, hardness, fusibility, fixedness, and change of color upon a slight touch of mercury, etc." Locke calls this a "complex idea" because it consists of a number of components, all joined together in a description. However, the real essence of gold must be something that underpins these characteristics. Locke conjectured that this must lie in the microscopic structure of gold.

> This essence, from which all these properties flow, when I enquire into it and search after it, I plainly perceive that I cannot discover: the furthest I can go is only to presume, that it being nothing but body, its real essence, or internal constitution, on which these qualities depend, can be nothing but the figure, size, and connection of its solid parts; of neither of which having any distinct perception at all, can I have any idea of its essence. . . . [51]

Today, we can marvel at Locke's prescience. Although he couldn't have known that the stuff his ring was made from had atomic number 79, he knew that there must be something about its microscopic structure that made it gold.

Like the alchemists before him, Locke applied the same pattern of reasoning to human beings. To demonstrate this, he asks us to consider a child who has "framed the idea of a man." Locke suggests that "it is probable that his idea is just like that picture which the painter makes of the visible appearances joined together"—that is, the child forms a complex idea of what a man is on the basis of the characteristics of the people whom he or she has observed. Locke then asks us to suppose that the child is English, and is exposed only to people who are "white or flesh-colour." It follows that "the child can demonstrate to you that a negro is not a man, because white colour was one of the constant simple ideas of the complex idea he calls man; and therefore he can demonstrate . . . that a negro is not a man." Locke continues, "And to this child, or any one who hath such an idea . . . can you never demonstrate that a man hath a soul, because his idea of man includes no such notion or idea in it."[52] The child's way of thinking misses the mark, because it fails to zero in on the real essence of the human.

Locke thought that even though real essences are impossible to observe, we suppose that they exist.

> First, it is usual for men to make the names of substances stand for things, as supposed to have certain real essences, whereby they are of this or of that species. . . . Who is there almost, who would not take it amiss, if it should be doubted whether he called himself a man, with any other meaning, than of having the real essence of a man? And yet if you demand what those real essences are, it is plain men are ignorant and know them not. . . . And yet, though we know nothing of these real essences, there is nothing more ordinary than that men should attribute the sorts of things to such essences.[53]

This sort of essentialistic thinking can lead us astray. Although it works nicely for chemical elements like gold and silver, it doesn't apply to biological species. Biological species do not have essences—at least not in the traditional sense.[54] There is no hidden essence unique to North American porcupines, nor is there for *Homo sapiens*. But Locke was right to suspect that essentialism comes naturally to us.

Locke's theory of real essence was dramatically extended by Princeton University philosopher Saul Kripke. Although not well known outside of the circle of academic philosophers, Kripke is one of the most influential philosophers of recent times. One of his most important contributions has been to clarify the role that essences play in our talk about natural kinds. Kripke illustrated his ideas using the Lockean example of gold.

> Given that gold does have the atomic number 79, could something be gold without having the atomic number 79? . . . Given that gold is this element, any other substance, even though it looks like gold and is found in the very places where we in fact find gold, would not be gold. It would be some other substance which was a counterfeit for gold.[55]

The idea here is disarmingly simple. The word *gold* is reserved for the stuff with atomic number 79. Anything that looked and behaved exactly like gold, but didn't have atomic number 79 wouldn't be gold any more than a brush-stroke for brush-stroke copy of the *Mona Lisa* would be the *Mona Lisa*. Both would be mere simulations of the real thing.

Locke seemed to think that we have a natural tendency to essentialize human beings, thinking of them—or rather, of *us*—as members of a natural kind with a distinctive human essence. He was right. There is a substantial body of psychological research (some of which I will describe in Chapter Six) showing that human beings are natural-born essentializers. We spontaneously divide the world into natural kinds to which we attribute hidden essences. In doing so we suppose that there is a natural kind to which *we* belong—a *human kind*—and it's an aspect of our basic psychological makeup to think that the term *human* (or its equivalent) is properly applied only to bearers of a human essence.

This feature of human psychology opens up the possibility of someone appearing human without really being human. It's easy to imagine that a being can seem human without being human. The notion of demonic possession is an evocative example. To conceive of demonic possession you have to find it credible that a nonhuman spirit can inhabit a human body, and therefore that someone can be outwardly human, but inwardly demonic. Of course, belief in demons is a primitive superstition, but it's nevertheless a very compelling one, as is evidenced by the success of films like *The Exorcist*, one of the highest-grossing (and scariest) films ever made. The point is that we have no difficulty conceiving of demonic possession. In fact, this belief has been (and is) extremely widespread—presumably, because there is something about it that the human psyche finds congenial.

Zombies are another case in point. They are animate human bodies without souls—shells of human beings, with nothing inside, just like the lifelike robots that are standard fare in science fiction. The Terminator is easy to mistake for a person, but he's not a person. He's outwardly, but not inwardly, human. In fact, Australian philosopher David Chalmers is famous for his argument that we can conceive of beings that resemble people down to the last physical detail, and behave just like people do, even though they are completely devoid of consciousness. Even though these hypothetical beings, called "philosophical zombies," are different from those featured in Haitian folklore and horror films, the fact that Chalmers's theory is given credence by a segment of the professional philosophical community indicates that even highly educated people find it easy to embrace the metaphysical presumption underpinning the psychology of dehumanization.[56]

In this book, I will argue that when we dehumanize people we think of them as *counterfeit human beings*—creatures that look like humans, but who are not endowed with a human essence—and that this is possible only because of our natural tendency to think that there are essence-based natural kinds. This way of thinking doesn't come from "outside." We neither absorb it from our culture, nor learn it from observation. Rather, it seems to reflect our cognitive architecture—the evolved design of the human psyche.

The notion that a creature can appear human without being human invites a comparison with counterfeit gold. To try this on for size, let's look at the passage from Kripke again, this time substituting the term *human beings* for "gold," and *human essence* for "atomic number 79."

Given that human beings do have a human essence, could something be human without having this essence? . . . Given that humans are this kind of being, any other being, even though it looks human and is found in the very places where we in fact find human beings, would not be human.

It would be some other creature which was a counterfeit human.

I think that this reworded passage captures what goes on in the minds of people when they regard their fellow human beings as less than human. In the chapters to follow, I will add a considerable detail to this bare sketch as well as trace out some of its implications.

4

THE RHETORIC OF ENMITY

Since propaganda as the rhetoric of enmity aims to persuade people to kill other people, others must be demonized in a denial that we share a common humanity.
—NICHOLAS JACKSON O'SHAUGHNESSY,
POLITICS AND PROPAGANDA[1]

WE ARE MYTHMAKERS who weave stories to explain who we are, our origin and our destiny. Narratives about quasihuman monsters and demons with uncanny powers grip us like no other. These stories may be elegantly wrapped in the jargon of science or religion, culture or philosophy—but in the final analysis they are offspring of the human imagination against which rationality offers us only meager protection.

The Jewish-German philosopher Ernst Cassirer understood this very well. Writing in the aftermath of World War II, and desperately trying to comprehend the roots of the madness that had apparently seized his countrymen, Cassirer warned that

In all critical moments of man's social life, the rational forces that resist the rise of the old mythical conceptions are no longer sure of themselves. In these moments the time for myth has come again. For myth has not been really vanquished and subjugated. It is always there, lurking in the dark and waiting for its hour and opportunity. The hour comes as soon as the other binding forces of man's social life, for one reason or another, lose their strength and are no longer able to combat the demonic mythical powers.[1]

Cassirer's characterization of the Nazi phenomenon was far closer to the truth than Hannah Arendt's undeservedly reiterated comment about the banality of evil. Arendt used this phrase to describe Adolf Eichmann, the man in charge of the logistics of the Final Solution, whom she held up as the prototypical Nazi bureaucrat. However, this slogan was singularly inapplicable to Eichmann, a deeply committed anti-Semite who once remarked "I shall laugh when I jump into the grave because of the feeling that I killed five million Jews. That gives me a lot of satisfaction and pleasure."[3] Eichmann and his colleagues weren't moved by pale abstractions. Contrary to popular myth, genocide is never inspired by the thought that human beings are numbers, abstractions, or products on an assembly line. There was nothing banal about the narrative images at the heart of the Nazi project. They were dramatic, vivid, and apocalyptic. They were stories of salvation and destruction, of bloodsucking Jews defiling Aryan purity with their filth and corruption. More concretely still, they were stories of rats and lice, of blood-born infection, disease, and decay. These are images that strike a deep chord in the human psyche, for reasons that we will shortly be in a position to explain.

SHACKLED TO THE GREAT CHAIN OF BEING

> What we often call the beginnings of human history are also the beginnings of bondage.
> —KEVIN BALES, *DISPOSABLE PEOPLE: NEW SLAVERY IN THE GLOBAL ECONOMY*[4]

In 1862, Frederick Law Olmsted (the designer of New York's Central Park, as well as many other American landmarks) published a book entitled *The Cotton Kingdom: A Traveller's Observations on Cotton and Slavery in the American Slave States,* which described his travels through the Southern states in the decade just before the Civil War. At one point in the book, Olmsted recounts an exchange with a plantation overseer—a man whose job it was to force slaves to work. Olmsted inquired of the man if he found it "disagreeable" to whip the slaves. "I think nothing of it," he drawled in response. "Why, sir, I wouldn't mind killing a nigger more than I would a dog."[5]

Shortly afterward, while riding together across the plantation, the overseer spotted a black teenage girl hiding in the underbrush. After a few questions, he concluded that she was trying to avoid work. Dismounting from his horse, the overseer ordered her to get down on her knees. As Olmsted looked on:

> The girl knelt on the ground; he got off his horse, and holding him with his left hand, struck her thirty or forty blows across the shoulders with his tough, flexible, "raw-hide" whip. . . . At every stroke the girl winced and exclaimed, "Yes, sir!" or "Ah, Sir!" or "Please, Sir!" not groaning or screaming.
>
> He paused for a moment and ordered the young woman to tell him the truth.
>
> "You have not got enough yet," said he; "pull up your clothes—lie down." The girl . . . drew closely all her garments under her shoulders, and lay down upon the ground with her face toward the overseer, who continued to flog her with the raw-hide, across her naked loins and thighs, with as much strength as before. She now shrunk away from him, not rising, but writhing, groveling, and screaming, "Oh, don't, sir! oh, please stop, master! please, sir! please, sir! oh, that's enough, master! oh, Lord! oh, master, master! oh, God, master, do stop! oh, God, master! oh, God, master!"

Olmsted rode on alone. "The screaming yells and the whip strokes had ceased when I reached the top of the bank," he wrote. "Choking, sobbing, spasmodic groans only were heard." The slave driver soon caught up with him, laughing, and said, "She meant to cheat me out of a day's work, and she has done it, too."[6]

Slavery is as old as civilization, and has been practiced all over the world. It was ubiquitous in antiquity, and is taken for granted in both the Old and New Testaments of the Bible (you may recall that Paul enjoined slaves to obey their masters "in fear and trembling" as they would Christ), as well as in the Koran.[7] It has proven to be so robust, so resilient to suppression, that even today there are many millions of men, women, and children who are slaves. The total number, worldwide, is estimated as somewhere between 12 and 30 million.

The institution of slavery seems to have begun around ten thousand years ago, when the discovery of agriculture led nomadic tribes to settle down and till the soil. For the first time, wealth and property became important features of social life, and the egalitarianism of hunter-gatherer society gave way to rigid and oppressive systems of social stratification. Populations swelled, as a large labor force was required to till and harvest, to construct and maintain settlements, and to build fortifications against the depredations of others. Civilization building required the muscle power of countless human and nonhuman animals. Roughly two thousand years prior to the invention of agriculture, nomadic tribes had begun to domesticate animals, starting with the dogs that helped them hunt game, and moving on to animals like sheep and goats, which, in a world without refrigeration, served as walking larders. With the advent of agriculture, domesticated animals became the first farm machines—the sinews of oxen and donkeys supplemented those of human laborers in the backbreaking work demanded by the new soil-based economies.

This was when human beings began waging wars of conquest. Kings and emperors, believed to be earthly avatars of divine beings, had armies at their command for robbing others of the fruits of their labor. These bloody exploits, which were aided by the manufacture of weapons specifically designed to kill humans, gave rise to a new method for replenishing and expanding the labor force. Instead of slaughtering all of their enemies, the victors brought some of them back as slaves. These human beasts of burden were recruited to work alongside oxen and donkeys, or, in the case of many captive women, to satisfy the victors' sexual urges. The origin of slavery in warfare is preserved in the etymology of the word *servant*, which comes from the Latin *servare* ("save"). Servants were "saved" for forced labor instead of being summarily executed.[8]

There are unmistakable parallels between the treatment of slaves and the treatment of domestic animals. Brown University historian Karl Jacoby points out that that virtually all of the practices deployed for controlling livestock—practices such as "whipping, chaining, branding, castration, cropping ears"—have also been used to control slaves.

> In medieval Europe, a new slave would place his head under his master's arm, and have a strap placed around his neck, in imitation of a sheep or cow, and in eighteenth-century Britain, goldsmiths advertised silver padlocks "For blacks or dogs."

But slavery confronted slave owners with a moral problem. How could the obvious humanity of slaves be reconciled with their status as livestock? "Slavery was an institution that treated humans like domestic animals," writes Jacoby. "Yet clearly humans and livestock were not the same—or were they? The easiest solution . . . was to invent a lesser category of humans that supposedly differed little from brute beasts." In light of this, it's not surprising to learn that in ancient times slaves (and, equivalently, prisoners of war) were often sacrificed as offerings to the gods, along with sheep, cattle, and other kinds of livestock.[9]

Because slaves were spoils of war, they usually spoke a different language than that of their masters. Jacoby argues that this contributed to their dehumanization. "As the ability to communicate through speech is one of the most commonly made distinctions between humans and animals," he remarks, "the captive's lack of intelligible speech . . . most likely made them appear less than fully human."[10]

Jacoby's conjecture has an ancient pedigree. I mentioned in Chapter Two that the ancient Greeks divided people into two categories: themselves and barbarians. Over two thousand years ago, the Greek writer

Strabo suggested that the word *barbarian* literally meant "the bar bar people"—people who utter sounds like "bar, bar" instead of intelligible speech (a contemporary American might say "the blah, blah people").[11] On this basis, classical scholar John Heath argues,

> The barbarian Other . . . is primarily language deficient. . . . The close connection between speech and reason made it easy to assume that barbaroi —like slaves (and animals, I would add)—lacked both. With the loss of speech and reason, foreigners were in danger of losing all of their humanity.[12]

As intriguing as this theory is, it's just not credible. The ancient world was a sea of linguistic diversity, so it's ridiculous to think that a maritime trading people like the Greeks believed that those who spoke a foreign tongue had no language at all. Given their ethnocentric bias, it's far more likely that the Greeks simply looked down upon and ridiculed the languages spoken by foreigners. The problem wasn't that barbarians had no language: it was that their languages were inferior. Sneering references to the speech of a despised group is an effect, rather than a cause, of their dehumanization—and not just for the ancient Greeks. It is common for xenophobes to claim that the "lower" races are by their very nature incapable of mastering the "higher" languages. For example, German nationalists alleged that Jews were unable to master the German language, and thus resorted to the mongrel jargon of Yiddish (in one far-right tract published in the United States during the 1940s, Jews are described as uttering animal sounds— "wailing yelps and weird wails" and making "screech" noises—rather than communicating in a human fashion).[13]

There is ample evidence for rampant xenophobia in the ancient world. The ancient Egyptians referred to themselves, in grand ethnocentric style, as the "human beings" (*remtu*), implying that all others are nonhuman, a point which is made painfully explicit in texts like *The Instruction of Ani*, written around 1200 BCE, where the arduous task of teaching foreigners to speak Egyptian is compared to the practice of training nonhuman animals.

The savage lion abandons his wrath, and comes to resemble
the timid donkey. The horse slips into his harness, obedient
it goes outdoors. The dog obeys the word, and walks behind
its master. . . . One teaches the Nubian to speak Egyptian,
the Syrian and other foreigners too.[14]

In a typically grandiose description of the military exploits of Phar-
aoh Amenemhet I, who ruled Egypt from 1985 to 1956 BCE, the enemies of Egypt are represented as nonhuman predators. "I subdued
lions, I captured crocodiles," he boasted. "I repressed those of Wawat, I
captured the Medjai, I made the Asiatics do the dog walk."[15]

The Mesopotamians also denied the humanity of their neighbors.
A three-thousand-year-old Babylonian text entitled *The Curse of Agade*
presented the Gutians as "not classed among people . . . with human
instinct but canine intelligence and monkeys' features," and elsewhere
described them as "serpents of the mountains" and "dogs." The Amor-
ites were said to have "instincts like dogs or wolves," and other groups
were said to have "partridge bodies and raven faces." Archaeologists believe that the images of warriors subduing wild animals that adorn
cylinder seals from the fourth millennium BCE may depict Mesopotamian troops subduing their neighbors.[16]

Further east, the Chinese distinguished themselves from border and
tribal people by assigning them labels with dehumanizing implications.
The barbarian tribes living to the east were called *"I-ti,"* a name containing the Chinese character *ch'uan*, which means "dog," and the barbarian tribes to the north were *"I-man,"* which includes the character *ch'ung*,
meaning "insect." The Rong-Di were called "wolves." In the *Shanhai-
jing* or *Classic of Mountains and Seas*, an account of the geography of
ancient China dating from the fourth century BCE, only the Chinese
are referred to as human beings *(ren)*. The distinguished sinologist E.
G. Pulleyblank points out that prior to the middle of the eighth century
BCE these barbarian groups were "looked on as not quite human."[17]

These facts point to the likelihood that foreigners were dehumanized
before being enslaved, and that slavery merely reinforced their subhuman status. To understand this more deeply, it's essential to consider

exactly what slavery is. We tend to think of slaves as people who are owned, and who can be bought and sold. But consider the fact that in today's world, professional athletes are regularly bought and sold, and yet they are not slaves. "While the terms of the transaction differ," remarks Harvard sociologist Orlando Patterson in his classic *Slavery and Social Death*, "there is no substantive difference in the sale of a football idol . . . by his proprietors . . . and the sale of a slave by one proprietor to another."[18] Not convinced? What about the reserve clause in professional sports, which made it possible for owners to buy and sell athletes against their will? Patterson reminds us: "Many sportswriters directly compare the reserve clause to slavery, Alex Ben Block's comment on the issue being typical: 'After the Civil War settled the slavery issue, owning a ball club is the closest one could come to owning a plantation.'" The retain-and-transfer system in British soccer was similarly compared with slavery.[19]

Patterson argues that the defining characteristic of slavery is not ownership but *social death*. The slave is a nonperson. For example, among the Cherokee:

> The slave acquired the same cultural significance . . . as the bear—a four-footed animal which nonetheless had the human habit of standing on its hind legs and grasping with its two front paws. . . . Similarly, the *atsi nahsa'i*, or slaves, were utterly anomalous; they had the shape of human beings, but no human essence whatever.[20]

The Greeks and Romans thought of their slaves as livestock. According to Notre Dame classicist Keith Bradley:

> The ease of association between slave and animal . . . was a staple aspect of ancient mentality, and one that stretched back to a very early period: the common Greek term for "slave," *andrapodon*, "man-footed creature," was built on the foundation of a common term for cattle, namely, *tetrapodon*, "four-footed creature."[21]

The principle of equating slaves with domestic animals was upheld by the Romans, who enshrined it in law. The *Lex Aquilia* statute of the third century BCE stated, "If anyone shall have unlawfully killed a male or female slave belonging to another or a four-footed animal, whatever may be the highest value of that in that year, so much money is he to be condemned to give to the owner." Centuries later, the jurist Gaius commented that under *Lex Aquilia* slaves were legally equivalent to domesticated animals, and Varro wrote in the late 30s BCE that some divide "the means by which the land is tilled" into two parts, namely "men and those aids to men without which they cannot cultivate." Included in the latter category are "the class of instruments which is articulate, the inarticulate, and the mute; the articulate comprising the slaves, the inarticulate comprising the cattle, and the mute comprising the vehicles."[22] Later, in the early centuries of the Islamic empire, slaves were collectively referred to as "heads" (*ru'us raqiq*)—as in "heads of cattle"—and slave traders were called "cattle dealers" (*nakhkhs*).[23]

A thousand years afterward little had changed, as is evidenced by Frederick Douglass's account of how, as an eleven-year-old boy, he and other slaves were treated as livestock in the estate of their diseased master.

I was immediately sent for, to be valued with the other property. Here again my feelings rose up in detestation of slavery. I had now a new conception of my degraded condition. . . . We were all ranked together at the valuation. Men and women, old and young, married and single, were ranked with horses, sheep, and swine. There were horses and men, cattle and women, pigs and children, all holding the same rank in the scale of being, and were all subjected to the same narrow examination. Silvery-headed age and sprightly youth, maids and matrons, had to undergo the same indelicate inspection.[24]

DEHUMANIZATION IN BLACK AND WHITE

This man born in degradation, the stranger brought by
slavery into our midst, is hardly recognized as sharing
the common features of humanity. His face appears to
us hideous, his intelligence limited, and his tastes low;
we almost take him for some being intermediate between man and beast.

—ALEXIS DE TOCQUEVILLE, *DEMOCRACY IN AMERICA*[25]

It is sometimes claimed that racism is a modern European invention,
no more than a few centuries old, and that it played no role in premodern and nonWestern forms of slavery. Neither of these views is correct.
Although premodern and non-Western people did not deploy the same
racial constructs as those that continue to haunt us today, the essence of
racism—the notion that whole populations possess an irredeemably
defective character—has been part of human culture for a very long
time. It is exemplified by the Greek attitude toward barbarians, and the
attitude of the ancient Chinese toward the frontier tribes surrounding
them. The ancient Egyptians explicitly distinguished between four
races: the Libyans, Nubians, Egyptians, and Semitic inhabitants of Palestine and Syria (the latter were especially despised, and frequently
disparaged as "wretched Asiatics") and they distinguished between
these groups in part by the color of their skin. The famous Roman physician Galen specifically claimed that black people have defective
brains, which make them unintelligent, a claim that was repeated by
the famous tenth-century Muslim writer al-Ma'sūdī.[26]

Racism is distinct from slavery, although the two often intersect. In
most slaveholding societies, both masters and slaves belong to the same
ethnic group, but in about 20 percent of slaveholding societies' masters
and slaves are divided along racial or ethnic lines. Patterson remarks
that, given these facts, the claim that American slavery was uniquely
racialized "betrays an appalling ignorance of the comparative data on
slave societies," and he takes issue with Frank Snowden's frequently re-

peated claim that there was no prejudice against black people in the Greco-Roman world.[27]

Discrimination against and oppression of black people is the form of racism most familiar to Western readers, who typically think of the transatlantic slave trade in this connection. However, racial discrimination against sub-Saharan Africans, and their dehumanization, has a much longer history. Prejudice against blacks seems to extend back to the earliest days of Islam, for Muhammad reportedly found it necessary to insist in his final sermon that "No Arab has any priority over a non-Arab, and no white over a black except in righteousness," and not long after, the black poet Subhaym, a slave who died in 660, complained that "the lord has marred me with blackness." There is clear evidence that, by the Middle Ages if not before, derogatory attitudes had tipped over into dehumanization. The well-known eleventh century Arab writer Sa'adi al-Andalusi, a judge in Islamic Spain, held that "the rabble of Bujja, the savages of Ghana, [and] the scum of Zanj," all of whom were sub-Saharan African peoples, "diverge from [the] . . . human order," and three centuries later, the great Tunisian scholar Ibn Khaldun informed his readers "the Negro nations . . . have attributes that are quite similar to those of dumb animals."[28]

Although the Arabs kept white as well as black slaves (the word *slave* comes from *Slav*, because many Arab slaves were Slavic people), the derogatory stereotypes associated with black people centuries later in the United States were already prevalent in medieval Islamic culture. Sub-Saharan Africans were seen as stupid, oversexed, lazy, dishonorable, coarse, and dirty. Ibn Butlan, an eleventh-century Christian physician from Syria who specialized in detecting concealed physical defects of slaves on behalf of potential purchasers, expressed the then current stereotype in his handbook for slave buyers when he described Zanj (East African) women in the following terms: "the blacker they are the uglier their faces. . . . there is no pleasure to be got from them, because of the smell of their armpits and the coarseness of their bodies." It is important to note that the contempt in which Arabs held many of their southern neighbors was not always attributed specifically to their *color*. Abyssinian and Nubian slave girls were highly prized by wealthy Arabs.[29]

The prevalence of these attitudes should come as no surprise. Although the form of slavery practiced by the Arabs was, on the whole, less cruel and degrading than its later American counterpart, it was nevertheless brutal and demeaning, as is evidenced by the fact that caravans that transported captive men and women across the Sahara to the Muslim kingdoms to the north were even more lethal than the densely packed ships that carried Africans across the Middle Passage to the Americas (Patterson notes that the mortality rate en route was between 3 and 7 percentage points greater than that of the transatlantic trade). It is difficult to treat humans so inhumanely while continuing to acknowledge their humanity.[30]

Now, let's consider the American experience. The transatlantic slave trade was pioneered in the fifteenth century by the Portuguese and Spanish, and gradually adopted by the British and others to supply labor for their burgeoning colonies. The ethnic prejudices that made the trade possible found support in scholarly speculations about Africans' rank on the great chain of being. Men of letters such as the physician Sir Richard Blackmore deployed the full weight of academic authority to support the claim that black Africans supplied the missing link between human beings and chimpanzees.

> As Man, who approaches nearest to the lowest Class of Celestial Spirits . . . so the Ape or Monkey, which bears the greatest Similitude to Man, is the next Order of Animals below him. Nor is the Disagreement between the basest Individuals of our Species and the Ape or Monkey so great, but that the latter were endow'd with the Faculty of Speech, they might perhaps as justly claim the Rank and Dignity of the Human Race, as the savage Hottentot or stupid Native of *Nova Zembla*.[31]

Although Blackmore stopped just short of describing Africans as nonhuman creatures, others were less restrained. Slave owners and merchants had a vested interest in the subhuman status of Africans, for if Africans were lower animals, then it was right and proper to treat them as such.

Much of what we know about the dehumanization of slaves in the North American colonies comes from the writings of a missionary named Morgan Godwyn. Born in England in 1640, he followed in his father's footsteps and became an Anglican minister. In 1666, he made his way to Virginia, and then to Barbados to act as a clergyman to all-white parishioners. At the time, planters in both British Caribbean and North American colonies didn't allow slaves to be baptized or receive religious instruction. On Barbados, only a few Quakers took an interest in the slaves' spiritual well-being. Although Godwyn was strongly opposed to Quakerism, he was deeply influenced by the Quakers' commitment to social justice, and this determined the future course of his religious and literary career.

Godwyn was determined to bring the gospel to the oppressed peoples of the New World. After years of conflict with plantation owners on Barbados, he returned briefly to England, and then embarked on a ship bound for Virginia, where he launched a relentless literary campaign. Godwyn's central charge was that slaveholders justified their treatment of Africans on the grounds that they had no souls, and therefore were not human. The colonists, he averred, held a "disingenuous position" that "the Negros, though in their Figure they carry some resemblances of Manhood, yet are indeed no men" and advocated "Hellish Principles . . . that Negros are Creatures Destitute of Souls, to be ranked among Brute Beasts and treated accordingly."[32]

Godwyn was not exaggerating. In 1727, ten thousand copies of an address by the Bishop of London (a huge number for that era) were distributed to colonists, in which he beseeched colonists to consider Africans "not merely as slaves and on the same level with neighboring beasts, but as *men* slaves and *women* slaves, who have the same frame and faculties with yourselves, and souls capable of being made happy, and reason and understanding to receive instructions to it." The Irish philosopher and clergyman George Berkeley, who visited Rhode Island in 1730, en route to an ill-fated attempt to establish a university for colonists and Native Americans on Bermuda, complained to members of a New England missionary society that "our first planters" had "an irrational Contempt of the Blacks, as Creatures of another Species, who [had] no Right to be admitted to the Sacraments." Somewhat later, in the 1741 trial for a black man accused of plotting a slave uprising in New York, the prosecuting attorney proclaimed that most blacks were "degenerated and debased below the Dignity of Human Species . . . the Beasts of the People," while another assured his readers that "the Negroes . . . are the most stupid beastly race of animals in human shape" and proposed a taxonomy of five types of Africans consisting of "1st Negroes, 2d, Orang Outangs, 3d, Apes, 4d, Baboons, and 5th, Monkeys."[33]

It seems incredible that educated and cultivated men could believe that it was literally true that, unlike whites, black people did not have souls. Even Winthrop Jordan, an eminent authority on slavery in early America, expressed incredulity. He wrote, "American colonialists no more thought that Negroes were beasts; if they had really thought so they would have sternly punished miscegenation for what it would have been—buggery." And yet, "the charge that white men treated Negroes as beasts was entirely justified if not taken literally."[34]

Jordan's comments reflect a simplistic view of the psychology of dehumanization. Did the colonists really believe that black people were subhuman? It all depends on what you mean by "believe." The word *belief* covers a great deal of territory. Sometimes, we use it to refer to statements that we mouth to ourselves and to others, but which have no impact on how we live our lives. Belief in God often falls into this category. There are people—I think, many people—who claim to believe in the Christian (or Jewish, or Muslim, or . . .) God, but whose behavior remains unaffected by their professed convictions (they conduct their lives in ways that are indistinguishable from unbelievers). In such cases, it seems reasonable to say that they don't *really* believe in God, because real beliefs guide behavior. We can say of such people that although they don't really believe in God, they nevertheless believe that they believe in him. Likewise, incredible as it may sound, there are people who profess to believe that the material world is an illusion—that it is "all in the mind" (the philosopher George Berkeley, whom we met earlier in the present chapter, was a famous exponent of this position). But threaten to pour a cup of hot coffee in such a person's lap, and he or she will take defensive action, just like anyone else would. People like this believe that the material world is perfectly real—but they falsely believe that they don't believe it.³⁵

Bearing these points in mind, it would be an error to take slaveholders' statements about Africans at face value. Actions really do speak louder than words, and words shouldn't be trusted unless they're backed up by behavior. If you want to know what people believe, look at what they do.

Many colonists both treated slaves as less than human and also explicitly stated that Africans were soulless animals. They justified their position on religious grounds. Some asserted that black people were not the progeny of Adam and Eve, but were descendants of creatures that were created before the first humans or were formed from debris left by the great flood (the same reasoning that was used to impugn the humanity of American Indians). Another, highly influential line of biblical exegesis drew on the account of Noah's curse on the descendants of Ham described in Chapter Nine of the book of Genesis. In this story, Noah planted a vineyard, used the grapes to brew his own wine, and became so inebriated that he passed out in his tent. Noah's son Ham entered the tent, and saw Noah sprawled out nude in a drunken stupor. He informed his two brothers, Shem and Japheth, who walked backward into the tent (to avoid laying eyes on their naked father) and covered him up. The scripture records that when Noah came to, he "knew what his younger son had done unto him" and cursed Ham's son Canaan, saying "a servant of servants shall he be unto his brethren."³⁶ If the punishment seems rather stiff for such a minor offense, notice the curious turn of phrase: Noah "knew what his younger son had done to him." What exactly had Ham done? According to some rabbinical commentators, the reference to Ham's "seeing" his father was a euphemism for raping or castrating him, which would explain why Noah was so miffed. Whatever the explanation, for centuries, Jewish, Christian, and Muslim scholars held that the name *Ham* was derived from a Hebrew word meaning "black" and "hot," and this was taken to imply that Ham's descendants were black people from a hot place (sub-Saharan Africa), and that therefore (so the shaky inference runs) black Africans are destined for enslavement (in fact, the etymology of *Ham* remains mysterious, but it is almost certainly not derived from the Hebrew for "black" and "hot"). This ancient exegetical tradition found a distant echo in "son of Ham shows"—popular carnival attractions in which white men paid for the pleasure of hurling baseballs at the head of a black man.[37]

American colonists added a new wrinkle to the "son of Ham" theory. They argued that Noah's curse condemned Ham's descendants to subhumanity. "Because they are Black," wrote Godwyn, "therefore they are Cham's Seed; and for this [reason] under the Curse, and therefore no longer Men, but a kind of Brutes."[38] Godwyn dismissed these biblical justifications as absurd, and insisted that there is no reason to think of Africans as anything less than fully human. "Godwyn's proofs were largely commonsensical," writes Columbia University historian Alden T. Vaughan. "Africans, he pointed out, obviously have human shape and appearance; and although they are of a darker complexion than the English and most other Europeans, so are five-sixths of the world's people."

> Even if one were to grant (as he did not) that blackness was a deformity, Godwyn insisted that it was no more a sign of bestiality than was any mental or physical abnormality; people with dark pigmentation can reproduce themselves, in fact they often reproduce with whites—a sure sign of their humanity. Godwyn's position on color preferences was also refreshingly even-handed for a seventeenth-century Englishman. There is no universal standard of beauty, Godwyn reminded his readers: blacks favor their own color just as whites do, and if a subjective objection to skin colors were allowed to consign some people to the category of brutes, fair-skinned Europeans might someday find themselves so labeled. Moreover, after a few generations in hot climates, Goldwyn argued . . . even the English "become quite Black, at least very Duskie and Brown." Are they to be considered brutes rather than men?[39]

Godwyn drove his point home by arguing that the slaveholders' own practices tacitly acknowledged the human stature of black people. He

noted that some slaves were placed in positions of considerable responsibility (for instance, as overseers of other slaves), but "it would certainly be a pretty kind of Comical Frenzie to employ cattel about
business, and to constitute them Lieutenants, Overseers, and Gover-
nours," and also noted that the fact that white men used slave women
for their own sexual gratification would make them guilty of bestiality
if slaves were really just livestock.

The dehumanization of African Americans did not end with the
creation of the new nation in 1776, or with the abolition of slavery in
1865. Books and pamphlets published during the latter part of the nineteenth and early twentieth centuries continued to assert that they were
beasts. During the nineteenth century the new discipline of anthropology gave this racist ideology a veneer of scientific credibility. Some, like
the British surgeon Sir William Lawrence, the Harvard geologist Louis
Agassiz, and the Philadelphia physician Samuel George Morton were
polygenecists—people who believed that each race had evolved independently of the others—and therefore that black people were a separate species ("That the negro is more like a monkey than a European,"
wrote Lawrence, "cannot be denied as a general observation"). The
German anthropologist Theodor Waitz vividly described the conflu-
ence of the polygenecist mindset with exterminationist policy. Writing
in 1863 of those who regard the so-called "lower" races as subhuman
creatures, he remarked:

> If there be various species of mankind, there must be a nat-
> ural aristocracy among them, a dominant white species as
> opposed to the lower races who by their origin are destined
> to serve the nobility of mankind, and may be tamed, trained
> and used like domestic animals, or may, according to cir-
> cumstances, be fattened or used for physiological experi-
> ments without any compunction. To endeavor to lead them
> to a higher morality would be as foolish as to expect that
> lime trees would, by cultivation, bear peaches, or the mon-
> key would learn to speak by training. . . . All wars of exter-
> mination, whenever the lower species are in the way of the

white man, are then not only excusable, but fully justifiable,
since a physical existence only is destroyed, which, without
any capacity for a higher moral development, may be
doomed to extinction in order to afford space to higher organisms.[40]

And he went on to add, with delicious irony, that:

Such a theory has many advantages. . . . It flatters our self-
esteem by the specific excellence of our moral and intellectual endowment, and saves us the trouble of inquiring for
the causes of the differences existing in civilization. The
theory has thus obtained many adherents; whilst there
are some who consider this one of the reasons which render
the assumption of a specifically higher mental endowment
of the white race, improbable.[41]

Polygenecism was sometimes dubbed the "American school" because of its popularity in the United States, especially among apologists
for slavery. Many Americans fused the Bible with biology to underwrite
their racial beliefs in an even more outlandish manner than their seven-
teenth- and eighteenth-century predecessors had done. Some religious
polygenecists believed that Africans are not of Noah's lineage, but
were descended from animals that he took aboard the ark. Others believed that blacks were the progeny of the devil, or descendents of a
subhuman race that God had fashioned prior to his creation of Adam.
A book entitled *The Negro a Beast*, published at the turn of the twentieth century by the American Book and Bible House, offered a bizarre
fusion of scriptural exegesis and biological fantasy. Its anonymous author informed his readers that the serpent of Eden was really a black
man (obviously using the word *man* rather loosely) who had evolved
from an ape, whereas Adam and Eve were white people created in
God's own image.[42]

Waitz was a monogenecist—a person who held that human beings
are a single species with a common origin. Monogenecists were more

often opposed to slavery than their polygenecist colleagues. (Charles Darwin was a prominent monogenecist, as well as a passionate abolitionist.) However, it would be wrong to assume too tight a connection between views of human origins and views about slavery. There was an abundance of religious monogenecists who considered polygenecist views to be heretical, and clung to the time-honored "curse of Noah" theory.[43]

Beliefs like these fueled the continued violence routinely directed at African Americans during the century or so following the Civil War. The story of Ota Benga, a Batwa ("pigmy") tribesman, provides a heartbreaking illustration of the dehumanization of Africans during this period. Ota Benga lived with his wife and children in a village in the vast tract of land in central Africa then called the Congo Free State. King Leopold II of Belgium founded the Congo Free State, ostensibly to provide aid to the people living there. However, the Congo Free State was anything but free. Leopold ruthlessly exploited its land and its people, draining it of resources like rubber, copper, and ivory, and exterminating approximately eight million people in the process. The *Force Publique*—a corps of African mercenaries, enforced the reign of terror. They did their job with gratuitous cruelty. Men, women, and children who failed meet their quotas were flogged with hippopotamus-hide whips, or had their hands hacked off with a machete. The hands were then collected in baskets, and presented to colonial officials. One eyewitness reported, "A village which refused to provide rubber would be completely swept clean. . . . I saw soldier Molili, then guarding the village of Boyeka, take a big net, put ten arrested natives in it, attach big stones to the net, and make it tumble into the river. . . . Soldiers made young men rape or kill their own mothers and sisters."[44]

Ota Benga's village was one of those "swept clean" by the *Force Publique*, who murdered his wife and children and sold him to an African slave trader.

It was at this point that Samuel Phillips Verner entered the picture. Verner, a missionary and entrepreneur, was in Africa on a mission— but not a religious one. He had recently signed a contract to bring exotic specimens of humanity to St. Louis for a "human zoo" at the 1904

World's Fair. This was to be a grand ethnographic exhibit, giving visitors an opportunity to ogle at tribal people brought to Missouri from the four corners of the world. Even the old Apache warrior Goyathlay—better known by his Mexican nickname "Geronimo"—was going to be on display. Verner was shopping for Pygmies. When he discovered Ota Benga, he paid off the slave merchant and took the young man to the United States, along with seven other Batwa who agreed to join them.

When the fair was over, Verner returned them all to their homeland, and remained in Africa for a year and a half collecting artifacts and animal specimens. During this time, he and Ota Benga became friends. Benga accompanied Verner on his collecting adventures, and eventually asked to return with him to the United States. Verner consented. After a brief stint in New York's Museum of Natural History, Ota Benga was given a home at the newly opened Bronx Zoo, where he soon became an exhibit, sharing a cage with an orangutan. "Few expressed audible objection to the sight of a human being in a cage with monkeys as companions," *The New York Times* wrote the next day, "and there could be no doubt that to the majority the joint man-and-monkey exhibition was the most interesting sight in Bronx Park."[45]

Spokesmen for the African-American community protested. Rev. James H. Gordon pleaded, "Our race, we think, is depressed enough, without exhibiting one of us with the apes. We think we are worthy of being considered human beings, with souls," and a delegation of black clergymen led by Rev. R. S. MacArthur addressed the matter in a letter to the mayor of New York. "The person responsible for this exhibition," he wrote, "degrades himself as much as he does the African. Instead of making a beast of this little fellow, he should be put in school for the development of such powers as God gave to him. . . . We send our missionaries to Africa to Christianize the people, and then we bring one here to brutalize him."[46] These protests did not focus entirely on the racially degrading character of the exhibit. They were also concerned that the exhibit supported Darwinism, which, then as now, was anathema to many Christians. Buckling under the pressure of controversy, zoo authorities released Ota Benga from his cage, and allowed him to wander freely around the zoo, where jeering crowds pursued him. The *Times* reported:

> There were 40,000 visitors to the park on Sunday, nearly every man, woman and child of this crowd made for the monkey house to see the star attraction in the park, the wild man from Africa. They chased him about the grounds all day, howling, jeering, and yelling. Some of them poked him in the ribs, others tripped him up, all laughed at him.[47]

What happened next is not entirely clear. One sultry summer day, Ota Benga decided to undress. Apparently, keepers tried to force him back into his clothes, and he responded by threatening them with a knife. He was promptly transferred to the Howard Colored Orphan Asylum. Ota Benga declined Verner's offer to return him to Africa because, despite his bad experiences in New York, they were nowhere near as bad as the horrors unfolding in his homeland. After a sexual scandal involving a teenage girl, he was transferred to Long Island, and eventually to Lynchburg, Virginia, where he attended Lynchburg Seminary and was employed in a tobacco factory. Ten years after arriving in the United States, longing to return to his homeland but unable to afford a steamship ticket back, Ota Benga put a bullet through his heart.

MORAL DISENGAGEMENT

This is the barbecue we had last night. My picture is to the left with a cross over it. Your son, Joe
—INSCRIPTION ON POSTCARD DEPICTING
THE CHARRED REMAINS OF JESSE WASHINGTON,
A LYNCHED BLACK FARM WORKER, WACO, TEXAS, 1916[48]

Let's now return to the question raised earlier in this chapter. Why did the overseer think that killing a "nigger" was of no greater consequence than killing a dog?

The great eighteenth-century economist and philosopher Adam Smith, taking a cue from his friend and countryman David Hume, argued that morality is built into human nature. It flows from our natural emotional resonance with others. "How selfish soever man may be supposed," wrote Smith in the opening passage of his 1759 book *The Theory of Moral Sentiments*, "there are evidently some principles in his nature, which interest him in the fortune of others, and render their happiness necessary to him, though he derives nothing from it except the pleasure of seeing it."

> Of this kind is pity or compassion, the emotion which we feel for the misery of others, when we either see it, or are made to conceive it in a very lively manner. That we often derive sorrow from the sorrow of others, is a matter of fact too obvious to require any instances to prove it; for this sentiment, like all the other original passions of human nature, is by no means confined to the virtuous and humane, though they perhaps may feel it with the most exquisite sensibility. The greatest ruffian, the most hardened violator of the laws of society, is not altogether without it.[49]

Smith's point was that morality is, at its core, a matter of gut feeling rather than rules and precepts. We naturally resonate with the feelings of those around us. But it's important to bear in mind that we are not just creatures of feeling; we also make use of an impressive array of concepts.

Concepts are like boxes into which we sort our perceptions. Take visual perception. When we look around us, we don't see nondescript shapes and patches of color. Rather, we categorize these visual impressions as objects. Right now, as I gaze to my right I detect a vivid red shape in my visual field. I see this not as an oddly shaped configuration of color, but as a glass of merlot perched on the corner of my desk. To have this perception, my experience had to be diffracted through a conceptual prism—I need to have concepts like "wine," "drink," and "glass," each of which presupposes a grasp of many others. Without all of this, I would know that something is there, but I wouldn't know what that something is. Our concepts endow our experiences with form and meaning.

The same principle holds true in the social world. To recognize someone as a person—a fellow human being—you need to have the concept of a human being. And once you categorize someone as human, this has an impact on how you respond to him. Of course, other animals can identify members of their own species, too; otherwise they wouldn't be able to find mates. But there's no reason to think that the vast majority of nonhuman animals have notions of "us" and "them": they *react* to organisms as "other" without *conceiving of them* as such. (Chimpanzees appear to be an exception, as I will explain in Chapter Seven.)

The capacity for conceptual thought has given our species great behavioral flexibility. We aren't locked into rigid instinctual patterns the way other creatures are, because we're able to sculpt our lives by designing and redesigning the frameworks that we use to make sense of the world. We are, to a very significant degree, architects of our own destiny.

This isn't to say that human malleability is limitless—far from it. Like any other, evolution equipped our species with certain inclinations, preferences, and cognitive capacities. We naturally enjoy certain tastes but not others, are startled by loud noises, form groups with dominance hierarchies, and feel joy, anger, and disgust, and so on as the situation demands. *But the powerful intellect with which natural selection endowed us enabled our ancestors to make the momentous discovery that they could engineer their own behavior.* The device that they invented for this purpose is what we call *culture*—the complex systems of ideas, symbols, and practices that structure and regulate our lives.

Through culture, we defy nature. Although the body rebels against swallowing nasty-tasting medicine, recruiting the gag reflex installed in our mammalian ancestors to protect them from swallowing toxic substances, we can force ourselves to swallow it. No other animal, not even the clever chimpanzee, is able to swallow a substance to which it has an aversion. But we *Homo sapiens* can use the concept of "medicine" to overcome our biological impulses (as we do when we allow physicians to draw blood from our veins or slice into our bodies). Thanks to the potent force of culture, we can perform acts that are, on the face of it, irrelevant or contrary to our own biological interests: to embrace celibacy, to mortify the flesh, to build monuments and perform rituals, to create religious doctrines and political ideologies, to sacrifice others' lives, and to sacrifice our lives for others. These are all results of our expertise at self-engineering.

Adam Smith was right to say human beings are naturally empathic creatures. But he was also aware that moral feelings are anything but simple and straightforward. We don't just empathize with people in pain, and seethe with anger at those who inflict needless pain; our feelings hang on the coattails of our interpretations of what's going on. When it comes to emotion, concepts call the shots.

Thanks to our empathic nature, most of us find it difficult to do violence to others. These inhibitions account for the powerful social bonds that unite human communities and explain the extraordinary success story of our species. But this generates a puzzle. From time immemorial men have banded together to kill and enslave their neighbors, rape their women, take over their hunting grounds, drink their water, or grow crops in their fertile soil. British philosopher A. C. Grayling calls this the war/peace paradox. "An anthropologist from another planet would infer from a modern city that its human occupants are rational and peaceable beings who work together for the common good in wonderfully sensible ways," he notes. However:

> The alien anthropologist would be only half-right. Let the spaceship land on another day in another zone—a war zone— and the inference would be that mankind is lunatic, destructive and dangerous. . . . This, then, is the paradox: outside the peaceful and flourishing city is the barracks where the soldiers drill and the factory where the guns are made; far beyond the horizon is the smoke and din of battle, where humans kill each other in horrifying ways, ways that, in the past 100 years alone, have resulted in nearly 200 million violent deaths.[50]

Given the highly developed social and cooperative nature of our species, how do we manage to perform these acts of atrocity? An important piece of the answer is clear. It's by recruiting the power of our conceptual imagination to picture ethnic groups as nonhuman animals. It's by doing this that we're able to release destructive forces that are normally kept in check by fellow feeling.

This insight isn't original. Many scholars have remarked on the power of dehumanization to promote moral disengagement. One of the first was Herbert Kelman, whose work I briefly described in Chapter Two. Kelman was a survivor of the Holocaust, so he knew from bitter experience what happens when inhibitions against violence are lifted, and he wanted to understand the psychological and social mechanisms that cause this to happen.

He concluded that there are three crucial factors at work. One is *authorization*. When persons in positions of authority endorse acts of violence, the perpetrator is less inclined to feel personally responsible, and therefore less guilty in performing them.

Kelman's hypothesis was dramatically supported by the famous obedience experiments conducted by Stanley Milgram in 1961. Milgram, who was also a Jew, was impressed by reports on the trial of Adolf Eichmann, the man who was responsible for the transport of millions of Jews to death camps. Eichmann evaded capture after the war, and with the help of a network of Catholic priests, escaped to Argentina in 1950 by way of Italy. He was captured by Israeli security agents in 1960, and was taken to Jerusalem to stand trial. During the trial, Eichmann repeatedly justified his actions by saying that he had only been following orders, just as his colleagues in the Nuremberg trial had done fifteen years before.[51] Milgram was intrigued. How far could obedience go? To answer this question, he designed and implemented a series of experiments in which a subject was instructed to administer what he or she believed to be increasingly powerful electrical shocks to a person in an adjoining room (who was a confederate of the experimenter). As the voltage was increased, the "victim" would pound the walls and scream in pain, and if the subject showed reluctance to continue, they were first told "Please continue," and then, if they were still hesitant, "The experiment requires that you continue," "It is absolutely essential that you continue," and, finally, "You have no other choice, you must go on." Milgram found that 65 percent of his subjects could be induced to deliver the maximal voltage to an innocent, suffering subject. He concluded that:

> [O]rdinary people, simply doing their jobs, and without any particular hostility on their part, can become agents in a terrible destructive process. Moreover, even when the destructive effects of their work become patently clear, and they are asked to carry out actions incompatible with fundamental standards of morality, relatively few people have the resources needed to resist authority.[52]

Eight years later, U.S. headlines were ablaze with news of the My Lai massacre. Men from the U.S. Army's Charlie Company had mowed down more than five hundred Vietnamese civilians—mostly elderly men, women, and children—in the village of Son My. *Time* magazine reported that Sergeant Charles Hutto, one of the perpetrators of the carnage, told an army investigator, "It was murder. I wasn't happy about shooting all the people anyway. I didn't agree with all the killing, but we were doing it because we had been told."[53] These events inspired Kelman to conduct a survey of ordinary Americans, asking them what they would do if they were ordered to shoot all of the inhabitants of a village, including old men, women, and children. Would they comply with the order or refuse to obey? A shocking 51 percent of his respondents replied that they would follow the order, and only 33 percent said that they wouldn't.[54]

Unquestioning obedience paves the way for *routinization*. Killing and torture become a workaday activity, just a job. This helps overcome moral inhibitions in two ways. First, following a rigid routine eliminates any need for making decisions, and thus entertaining awkward moral questions. Second, focusing on programmatic minutiae of his job makes it easier for the perpetrator to hide from the meaning of his actions.

As important as authorization and routinization are, Kelman thought that they are not sufficient for overcoming moral resistance to cold-blooded violence. For this, dehumanization is required. "To the extent that the victims are dehumanized," he wrote, "principles of morality no longer apply to them and moral restraints are more readily overcome." If others are not really human, we can treat them as we like, or as we are instructed, without moral reservations getting in the way.[55]

The idea that dehumanization causes moral disengagement comes up again a few years later in a study by Stanford University psychologist Albert Bandura, who put Kelman's hypothesis to an empirical test. Bandura and his coworkers recruited college students to participate in the experiment. The experimental setting consisted of three cubicles, into which students were escorted three at a time. Each of the cubicles was equipped with an "aggression device"—a contraption that supposedly delivered an electric shock at ten levels of intensity. It was arranged that, once in the cubicle, the students would overhear someone explaining to a group of people in an adjoining room that they were participating in an experiment dealing with the effect of punishment on the quality of collective decision-making, that the participants were recruited from a variety of social backgrounds, and that each decision-making group would consist of three people with similar attributes. This was, of course, an elaborate ruse orchestrated for the benefit of the true experimental subjects—the students sitting in their cubicles eavesdropping on the briefing.

Next, each of the students was told that he or she would supervise a group of three decision-makers. If a member of their group proposed an effective decision, an amber light would flash in the cubicle, and no action would be required of the student. But if an ineffective decision was proposed a red light would flash, and the student would have to give the participants an electric shock. They could give the shock at any level they chose, from one (mild) to ten (painful). Of course, in reality no shocks were administered.

After this, the students were made to believe that they were privy to a conversation between the experimenter and his assistant, broadcast on an intercom system. After the click of a microphone switch, the experimenter announces that the experiment will soon begin, but he's immediately interrupted by the assistant, who asks him where some scoring forms are kept. The microphone is "accidentally" left on, and during the ensuing exchange, the experimenter describes the decision-making group to his assistant in one of three ways. They're either described as *humanized* (perceptive, understanding, and so on), *dehumanized* (animalistic, rotten), or described in a neutral manner. The episode concludes with the experimenter discovering that the microphone was left on, and hastily switching it off.

Bandura devised this experiment to investigate whether the dehumanizing descriptions would have any effect on the students' punitive behavior. The outcome was precisely what one might expect.

> Dehumanized performers were treated more than twice as punitively as those invested with human qualities and considerably more severely than the neutral group. . . . Subjects gradually increased their punitiveness toward dehumanized and neutral performers even in the face of evidence that weak shocks effectively improved performance and thus provided no justification for escalating aggression. By contrast, with humanized performers, subjects consistently adhered to mild punishment.[56]

When the shocks failed to improve performance, the students' behavior was even more disturbing. "Under dysfunctional feedback," he commented, "subjects suddenly escalated punitiveness toward dehumanized performers to near maximum intensities."[57]

Beginning with Hume's brief comments in the *An Enquiry Concerning the Principles of Morals*, the idea that dehumanization promotes moral disengagement has gained widespread acceptance. When a group of people is dehumanized, they become mere creatures to be managed, exploited, or disposed of, as the occasion demands. Throughout history, propagandists have exploited this to serve their political ends. There's no better way to promote a war than by portraying the enemy as a bloodthirsty beast that must be killed in self-defense. And there is no better

way to whip up enthusiasm for genocide than by representing the intended victims as vermin, parasites, or disease organisms that must
be exterminated for the purpose of hygiene.

The architecture of our minds makes us vulnerable to these forms of
persuasion. Images like these speak to something deep inside us. If you
still believe that you are the exception, and are immune from these
forces, I hope that by the end of this book you will have embraced a
more realistic assessment of your capacity for evil.

5

LEARNING FROM GENOCIDE

I am not one accepted in your parish.
Bleistein is my relative, and I share
the protozoic slime of Shylock, a page
in Stürmer, *and, underneath the cities,*
a billet somewhat lower than the rats.
Blood in the sewers, pieces of our flesh
float with odure on the Vistula.

—EMANUEL LITVINOFF,
"TO T. S. ELIOT"[1]

HAVE YOU EVER WONDERED what it would be like to participate in genocide? Have you imagined yourself as a guard at Auschwitz herding new arrivals from the train to the gas chambers, or as a Rwandan Hutu hacking men, women, and children to death with a machete? Try to do it, graphically and realistically. Was it difficult? Now try again, this time with the help of political scientist Daniel Goldhagen.

> You cut him. Then cut him again. Then cut him again and again. Think of listening to the person you are about to murder begging, crying for mercy, for her life. Think of hearing your victim's screams as you hack at or "cut" her and then cut her again, and again, and again, or the screams of a boy as you hack at his eight-year-old body.

Your imagination probably recoils from this exercise. Picturing it makes you feel sick. But this reassuringly natural response presents a conundrum, for genocide *does* occur. As Goldhagen goes on to remark, "the perpetrators [of genocide] do it, and hear it. And they do it with zeal, alacrity, and self-satisfaction, even enjoyment."[2]

NEITHER MONSTERS NOR MADMEN

What kinds of people willingly perform acts that most of us have difficulty even imagining? The knee-jerk answer is that they are monsters. Wasn't Hitler a monster? How about Stalin, Mao, or Saddam Hussein? Surely, they were all monsters! But what *are* monsters?

Monsters have always haunted the dark places of human imagination. The eerie images on cave walls in Spain and southern France painted by anonymous Stone Age artists include monstrous portraits. Interspersed with lifelike images of Pleistocene fauna we find strange figures that are neither clearly human nor unequivocally nonhuman. Some of them have human bodies with animal heads. Another displays a distorted human face with a strange, animalistic snout, and yet another human-looking figure has a snakelike appendage for a head. The most famous prehistoric monster portrait of all is at the Trois-Frères cave in southwestern France, which shows "some kind of half-man, half-beast creature, with staring, glowering eyes, powerful looking humanoid legs, and a weird head bearing horns or antlers." (Archaeologists have dubbed the Trois-Frères portrait "the sorcerer" on the assumption that it depicts a shaman dressed in animal skin performing a rite of hunting magic.)[3]

Monsters embody everything that we find dangerous and terrifying, and which we want to keep at arm's length. It is in this capacity that they play a prominent role in the religious mythology of the high civilizations of the ancient Middle East. In Egypt, the paradigmatic monster was Apep, an immense serpent that perpetually threatened to drag the ordered world into a seething abyss of chaos and mayhem, while Mesopotamian texts described a similarly awesome creature named Tiamat, whose coils covered thirty acres (the prototype for the biblical monster Leviathan).

While gigantic creatures like these represented the wild, uncontrollable, destructive forces of nature, smaller, anthropomorphic monsters embodied the predatory danger posed by other animals—especially the human ones. "Like most monsters worth the name before or since," writes anthropologist David Gilmore, "these smaller . . . beasties were wont to tear people apart and eat them." People believed themselves to be continuously stalked by ferocious specters; vampirelike, legions of succubae ripped open people's throats, drank their blood, and devoured their flesh. They dripped deadly poisons from their jagged maws; their razor-sharp claws were polluted and lethal.[4]

Interest in matters monstrous continued throughout classical antiquity. Greco-Roman monsters typically "combined human and animal traits in shocking or terrifying ways."[5] The Roman author Pliny the Elder included detailed information about monsters in his encyclopedic *Natural History*; creatures like *cynocephali*, creatures with human bodies and dogs' heads said to dwell in the remote mountains of India, and the amazing manticore, a creature with "a triple row of teeth like a comb, the face and ears of a man, gray eyes, a blood-red color, a lion's body, and inflicts stings with its tail like a scorpion."[6] Pliny's lurid account of the "monstrous races" is what prompted Augustine to emphasize that what makes someone human is something internal—the possession of a rational soul—rather than their outer bodily shell.

Concerns about monsters continued through the Middle Ages. During this period, the most important source of information about monsters was a book entitled *Liber Monstrorum de Diversis Generibus—The Book of Monsters of Various Kinds*—which was written some time between the ninth and eleventh centuries. It sets out a taxonomy consisting of human beings, lower animals, and monsters. *Liber Monstrorum* describes the monstrous as a category, sandwiched between man and beast; less than human, but more than animal. It's this combination of human and animal traits that gives them their uncanny quality. Some are born of monster parents, while others began life as humans, but degenerate into monsters in consequence of their wickedness. But whatever their genealogy, monsters are invariably evil, harboring malice toward humans and hatred toward the divine order.[7] Their eyes glow continuously. Their teeth, angled in their mouth, are lethal weapons, underscoring their predatory, nonhuman character. "Made as they were," the text concludes "the order of creation must keep them on the outside."[8]

Hitler, Stalin, and others who perform wicked acts were not monsters. Monsters don't exist. Like it or not, these men were human beings, as were all mass killers throughout history.

So, why do we call them monsters?

Croatian journalist and novelist Slovenka Drakulić offers an answer. Describing her experiences observing the trial of Balkan war criminals in The Hague, men like Ratko Mladic, who slaughtered nine thousand Bosnian Muslims at Srebrenica, and the infamous Slobodan Milosevic, Drakulić remarks:

> You sit in a courtroom watching a defendant day after day, and at first you wonder, as Primo Levi did, "If this is a man." No, this is not a man, it is all too easy to answer, but as the days pass you find the criminals become increasingly human. You watch their faces, ugly or pleasant, the way they yawn, take notes, scratch their heads or clean their nails, and you have to ask yourself: what if this is a man? The more you know them . . . the more you realize that war criminals might be ordinary people, the more afraid you become.
> Why?
> This is because the consequences are more serious than if they were monsters. If ordinary people committed war crimes, it means that any of us could commit them. Now you understand why it is so easy and comfortable to accept that war criminals are monsters.[9]

Calling these people "monsters" merely dehumanizes the dehumanizers, and thus becomes a symptom of the disease for which it purports to be the diagnosis. Calling people monsters is a way of whistling past the graveyard, a way of reassuring ourselves that *they* are so very different from *us*. It's not the "order of creation" that keeps them on the outside: It's you and me.

Another distancing tactic is to call these people "sick"—that is, mentally ill. Surely, one might think, anyone who performs such terrible, brutal acts, or orders them to be done, has got to be severely psychologically disturbed. How many times have you heard Hitler, Stalin, Pol Pot, Saddam, and even Osama bin Laden described as "madmen," "psychotics," or "psychopaths"? It's true that mental illness can explain bizarre and sometimes violent human behavior. However, these excursions into do-it-yourself psychiatry are usually ill conceived, because the sort of "sickness" ascribed to genocidal killers has nothing to do with malfunctions of the central nervous system. It's a secular euphemism for evil—a moral diagnosis dressed up as a medical one. The senior officials in Hitler's regime are often touted as paradigmatic examples of genocidal madmen. But what's the evidence for this? If anyone was in a position to know it was Leon Goldensohn, the American psychiatrist assigned to care for the Nazi war criminals at Nuremberg in 1946. But Goldensohn didn't confirm the popular prejudice. Instead, he found that, "With the exception of Rudolf Hess and in the later stages of the trials possibly Hans Frank, the defendants at Nuremberg were anything but mentally ill. Alas, most of them were all too normal. . . ."[10] The desire to exterminate the Jews wasn't the Nazis' problem—psychological or otherwise. It was their solution. The same is true of the contemporary poster children for moral insanity: Muslim "terrorists." Psychiatrist Marc Sageman interviewed almost two hundred jihadists and reports finding "no obvious mental health problems."[11] There is no evidence that the people who plan and implement mass killing are more prone to psychopathology than any person, picked at random, walking down Main Street USA.

This brings us back to our original question: What is it that enables some people to engage in genocidal killing? To answer it, we need to get a foothold in the genocidal mind-set. We need to get a sense of what sort of concerns lead otherwise ordinary people—people like you and me—to commit acts of barely imaginable atrocity.

GOEBBELS IN LODZ

One morning, as Gregor Samsa was waking up from anxious dreams, he discovered that in his bed he had been changed into a monstrous verminous bug.

— FRANZ KAFKA, *THE METAMORPHOSIS*[12]

In the fall of 1939, just after the German conquest of Poland, a film crew arrived in the Polish city of Lodz. Charged with the mission of shooting documentary footage for a film entitled *The Eternal Jew*, the crew worked under the personal supervision of Hitler's propaganda minister, Joseph Goebbels. Goebbels was a committed anti-Semite, who was responsible for the Kristallnacht pogrom during which German synagogues were burned, Jewish businesses destroyed, and tens of thousands of men, women, and children were deported to concentration camps. Now he was in Lodz, shooting what he called his "Jew film"—a film which, according to historian and Holocaust scholar Stig Hornshøj-Møller, may have been responsible for convincing Hitler of the need to annihilate the Jewish people

It is easy to imagine Goebbels as nothing more than a twisted, evil man—a monster, if you will—cold-bloodedly conspiring to wipe innocent people off of the face of the earth. But this would underestimate his complexity, and his humanity. Goebbels sincerely believed that Jews were dangerous subhuman creatures. In his mind, destroying them wasn't an act of cruelty: it was his moral duty.

Prior to flying from Berlin to Lodz to supervise the filming, Goebbels ordered the director, Fritz Hippler, to shoot some preliminary footage capturing "everything of Jewish ghetto life" including the practice of kosher ritual slaughter (*sechita*). Hippler showed the film to Goebbels on October 16. According to Hippler,

> Goebbels . . . wanted to show me how somebody with a proper attitude toward the Jewish question would react. Almost every close-up was accompanied by shouts of disgust

and loathing . . . at the ritual slaughter scenes he held his
hands over his face.[13]

These weren't just histrionics. Goebbels recorded in his diary on
October 17 that the scenes that Hippler showed him were "so dreadful
and brutal in their details that one's blood freezes. One pulls back in
horror at so much brutality. This Jewry must be exterminated."[14]

The theme continues in diary entries written after he joined the
film crew in Lodz on November 2, and accompanied the cinematogra-
phers into the heart of darkness. "These are no longer human beings,"
he wrote, "but animals. It is, therefore, also no humanitarian task, but a
task for the surgeon. One has got to cut here, and that most radically.
Or Europe will vanish one day due to the Jewish disease." The next day
he spoke with Hitler about it, and noted in his diary that, "The Jew is
garbage. Rather a clinical than a social matter." Oddly, the Nazis had
quite a different attitude toward *real* nonhuman animals. The long, grue-
some scene in *The Eternal Jew* depicting ritual slaughter of livestock was
bound to have been upsetting to Hitler, who was an ardent vegetarian,
antivivisectionist, and exponent of animal welfare.[15]

Other Germans made excursions to the ghetto as well. The state-
sponsored leisure service *Kraft durch Freude* (*Strength through Joy*) organized bus excursions so that soldiers could observe the subhumans at
close range. A report produced by the Polish government in exile describes these grotesque urban safaris.

> Every day large coaches came to the ghetto; they take the
> soldiers through as if it was a zoo. It is the thing to do to
> provoke the wild animals. Often soldiers strike out at passers
> by with long whips as they drive through. They go to the
> cemetery where they take pictures. They compel the fami-
> lies of the dead and the rabbis to interrupt the funeral and to
> pose in front of their lenses. They set up genre pictures (old
> Jew above the corpse of a young girl).[16]

At the time of Goebbels's visit, the future was bleak for the Jews of
Lodz. They had been harassed and abused, their businesses closed, their

property confiscated, and the four great synagogues of the city had been burned to the ground. Before long, most would perish in extermination camps or die of starvation. But all that Goebbels could see were vermin: carriers of the Jewish disease—a disease that would engulf the world unless it was obliterated. In the film's most notorious scenes, a swarm of rats appears on the screen, followed by scenes of rats emerging from sewers and infesting bags of grain, and then by shots of street life in Lodz, while the narrator explains:

> Wherever rats appear they bring ruin, by destroying man-kind's goods and foodstuffs. In this way, they spread disease, plague, leprosy, typhoid fever, cholera, dysentery, and so on. They are cunning, cowardly, and cruel, and are found mostly in large packs. Among the animals, they represent the rudiment of an insidious and underground destruction, just like the Jews among human beings.[17]

The Eternal Jew emphasizes the connection between Jews and filth, decay, and disease in every sector of cultural life. For instance, the narrator gravely states that this "race of parasites" has no feeling for the "purity and cleanliness" of the German idea of art. A "smell of foulness and disease" pervades Jewish art, which is "unnatural, grotesque, perverted, or pathological." However, the heroic National Socialist movement was determined to extirpate the pestilence. There were around 230,000 Jewish residents of Lodz when German troops marched into the city in September 1939. Six years later, when the Russians liberated it, fewer than nine hundred were left alive.

Although it may seem to be an oxymoron, the architects of the Final Solution conceived of their task primarily as a *moral* one. They seamlessly elided images of physical filth and disease with concepts of moral impurity. The Jews embodied evil, just as German civilization embodied moral purity. The Nazis' worldview combined elements of both physical and moral danger in a package not unlike contemporary Americans' image of the Islamic fundamentalist Taliban (Hitler often remarked that Jews "are very radical and have terroristic inclinations").[18] This apocalyptic vision is well summarized by Jay Gonen in his book on *The Roots of*

Nazi Psychology. Gonen writes that the Nazis believed that "we live in a polarized and most dangerous world in which nothing is safe. . . ."

> A pervasive danger floated over the polarized world. . . . Everything now seemed exposed to an insidious and corrupting agent of sickness. Life is dominated by the need to protect against the killing disease. . . . Hitler's perception was clear and ominous. An ill-understood evil is on the verge of triumph in this world. Nevertheless it can still be successfully fought. Total education could produce understanding of the disease while only total fanaticism could sustain the all-out combat necessary to stem its course and even to eradicate it once and for all.[19]

The stakes couldn't be higher. The fate of the world rested on the shoulders of those men and women with the sagacity to appreciate the Führer's message, and the courage to embrace their duty to humanity. There is no more revealing example of this peculiar moral sensibility than Nazi security chief Heinrich Himmler's speech to the SS officers in the Polish town of Poznan on October 4, 1943.

> I am referring to . . . the extermination of the Jewish people. . . . Most of you men know what it is like to see 100 corpses lying side by side, or 500 or 1,000. To have stood fast through this and—except for cases of human weakness—to have stayed decent, that has made us hard. . . .

He goes on to insist that all the wealth confiscated from murdered Jews was turned over to the Reich, and that SS did not keep any booty for themselves ("not . . . so much as a fur, as a watch . . . one mark or a cigarette"). Stealing confiscated goods is *immoral*, and those few who yielded to temptation would be executed.

> We had the moral right, we had the duty to our people to destroy this people. . . . We do not want, in the end, because

we have destroyed a bacillus, to be infected by this bacillus and to die. I will never stand by and watch while even a small rotten spot develops or takes hold. Wherever it may form we will together burn it away. All in all, however, we can say that we have carried out this most difficult of tasks in a spirit of love for our people. And we have suffered no harm to our inner being, our soul, our character.[20]

This theme is repeated in the letters and memoirs of patriotic Germans of the period, and not just the top brass. They took themselves to be mounting a defense of humanity against subhumanity, civilization against barbarism, virtue against moral degradation. One young serviceman wrote from the Eastern Front in 1941, where Russians were dying like flies and mobile killing units were slaughtering Jews en masse, that he was motivated by "the struggle for the truly human . . . the timeless cause of the spirit." Another wrote home that the war is all about "the preservation of human dignity, which is purified by pain and renunciation . . . the battle with the ghostly manifestations of Materialism." A company commander told his men that they were fighting to "revive the ancient virtues buried under layers of filth." Propaganda leaflets handed out to the troops assured them that "We would insult the animals if we described these mostly Jewish men as beasts. They are the embodiment of the Satanic and insane hatred against the whole of noble humanity . . . the rebellion of the subhumans against noble blood."[21]

It may seem barely comprehensible that intelligent men and women could view the world in such a hideously distorted way, and, turning their Manichean vision into policy, wage a genocidal war against a harmless people. It requires little effort to condemn the Nazis. Moral outrage comes cheaply. It is more difficult, and surely more valuable, to address those features of the human condition that precipitated the tragedy.

GENOCIDE

. . . Paragon of art,
That kills all forms of life and feeling
Save what is pure and will survive.

—ROY CAMPBELL, "AUTUMN"[22]

Many—perhaps most—researchers into genocide agree that efforts to destroy an entire people are almost always accompanied by the idea that those being annihilated aren't really people. As Daniel Goldhagen notes, "[T]he term *dehumanization* is rightly a commonplace of discussions of mass murder. It is used as a master category that describes the attitudes of killers, would-be killers, and larger groups towards actual and intended victims."[23]

Gregory H. Stanton, founder and president of the human rights organization Genocide Watch, describes dehumanization as a regular feature of genocides. "One group," he writes, "denies the humanity of the other group. Members of it are equated with animals, vermin, insects or diseases. Dehumanization overcomes the normal human revulsion against murder. At this stage, hate propaganda . . . is used to vilify the victim group." Likewise, University of Nebraska psychologist David Moshman notes that members of the victimized group "are construed as elements of a subhuman, nonhuman, or antihuman collective." Psychologist Clark McCauley and sociologist Daniel Chirot make a similar observation: "In most genocidal events the perpetrators devalue the humanity of their victims, often by referring to the victims as animals, diseased, or exceptionally filthy . . . notably pigs, rats, maggots, cockroaches, and other vermin" (or with "monstrous" creatures).[24]

With these observations in mind, let's briefly survey the role that dehumanization has played in some of the worst episodes of genocidal slaughter of the last hundred years.

When, during the late nineteenth and early twentieth centuries, the European powers were carving up the "dark continent" of Africa, they treated the indigenous people as beasts of burden, and sometimes as dangerous animals. "The blacks give an immense amount of trouble . . . ," complained explorer Henry Morton Stanley, an agent of the Belgian King Leopold II. "When mud and wet sapped the physical energy of the lazily-inclined, a dog-whip became their backs, restoring them to a sound—sometimes to an extravagant—activity."[25]

While the Belgians were helping themselves to the rich natural resources of the Congo, German colonists were busy plundering what is now the nation of Namibia, where they committed the first full-blown genocide of the twentieth century. The German and Dutch colonists regarded the indigenous Nama and Herero people with contempt, an attitude that they made no effort to conceal. A petition sent to the Colonial Department in Berlin by white farmers in the region stated, with alarming explicitness, that it is "impossible to regard them as human beings."[26] A contemporary missionary report confirmed that:

> [T]he average German looks upon and treats the natives as creatures being approximately on the same level as baboons (their favorite word to describe the natives). . . . Consequently the whites value their horses and oxen more highly than they do the natives. Such a mentality breeds harshness, deceit, exploitation, injustice, rape, and, not infrequently, murder as well.[27]

The abuses ignited a rebellion, which was brutally suppressed by General Lothar von Trotha, with the help of 14,000 troops imported for this purpose from the Fatherland. Von Trotha tailored his military strategy to the perceived status of the rebels. "Against nonhumans," he wrote, "one cannot conduct war humanely." Many were shot outright. Others were driven into the Kalahari Desert, where soldiers had poisoned the water holes. "Like a wounded beast the enemy was tracked down from one water hole to the next," states an official military report, "until finally he became the victim of his own environment." Some were burned alive. One eyewitness to the incineration of twenty-five men, women, and children reported, "The Germans said 'We should burn all these dogs and baboons in this fashion.'" Survivors were imprisoned in concentration camps. Around 60,000 Herero (around 75 percent of the original population), 10,000 Nama (half of the original population), and up to 250,000 others, men, women, and children, perished.[28]

Ten years later, genocide erupted in Turkey. Even before the blood-bath of 1915–16, Muslim Turks periodically butchered members of Christian minorities (primarily Armenians, but also Greeks and Assyrians) who were forced to endure the degrading status of dhimmitude. "The Turkish rule," wrote British ethnographer William M. Ramsey, "meant unutterable contempt. . . ."

> The Armenians (and the Greek) were dogs and pigs . . . to be spat upon, if their shadow darkened a Turk, to be outraged, to be the mats on which he wiped the mud from his feet. Conceive the inevitable result of centuries of slavery, of subjection to insult and scorn, centuries in which nothing that belonged to the Armenian, neither his property, his house, his life, his person, nor his family was sacred or safe from violence—capricious, unprovoked violence—to resist which by violence meant death![29]

The spirit of these pogroms, as well as their scope, is evident from the text of a letter sent by a Turkish officer to his parents and brother, and intercepted by British agents in 1895. "My brother," it begins, "if you want news from here we have killed 1,200 Armenians, all of them as food for the dogs. . . ." Between 1894 and 1896, as many as 100,000 Armenians were killed in clashes with government forces and Muslim militias. In the decade to follow, resurgent Turkish nationalism put further pressure on non-Muslim minorities. Threatening letters were sent to the Armenian press, one of which promised to "clean up the Armenian infidels who have become tubercular microbes for us."[30]

This promise was fulfilled in the spring of 1915. Plans for the genocide were finalized at a secret government meeting, where officials discussed a ten-point strategy, which included the decision to arrest and kill any Armenians who had ever worked against the government, to close all Armenian societies, and to stir up anti-Armenian feeling among Muslims to provoke organized massacres. Soon all Armenian men in the Turkish army were disarmed and massacred, and thousands of criminals were released from prison to form mobile killing units. The victims were either killed outright or forced to join a death march into the Syrian Desert, without food, water, or protection from the elements. The roads were littered with emaciated corpses. Most died of hunger, or were killed, along the way, and the remainder were exterminated after having reached their destination. As is usually the case in genocide, the rape of women and young girls was commonplace.

The Turkish authorities didn't characterize their victims as wild animals. They conceived of them in much the same way as the Nazis would later imagine Jews—as disease organisms infecting the body of the state—and announced the need to "rid ourselves of these Armenian parasites." Mehmed Resid, a professor of legal medicine at Istanbul Medical School and a major player in the genocide, described Armenians as "dangerous microbes" and asked, with grim rhetorical force, "Isn't it the duty of a doctor to destroy these microbes?" Armenians and other non-Muslim minorities were also identified with traditionally unclean animals such as rats, dogs, and pigs. As many as a million and a half men, women, and children were wiped out, usually by starvation, stabbing, clubbing, or asphyxiation, and also by burning and drowning. They were rarely shot, as bullets were deemed to be too valuable to waste on subhuman creatures.[31]

For most readers of this book, the word *genocide* is probably synonymous with Auschwitz. The Holocaust was the paradigmatic twentieth-century genocide, and is also the most thoroughly documented one. There is an immense literature describing how Germans of the Third Reich thought of Jews, as well as Slavs and Gypsies, as less than human, portraying them as apes, pigs, rats, worms, bacilli, and other nonhuman creatures. And it is abundantly clear from this evidence that the Nazis did not intend the term *subhuman* to be taken metaphorically. "One does not hunt rats with a revolver," quipped one SS expert, in a chilling allusion to the mass exterminations, "but with poison and gas."[32]

Perhaps the best way to get a sense of the centrality of dehumanization for the Nazi project is to look at its role in *Mein Kampf*, Hitler's
1925 autobiography-cum-ideological screed. After closely examining
the patterns of figurative language in *Mein Kampf*, German scholar
Andreas Musolff has concluded:

> The source imagery of Hitler's political worldview con-
> sisted in the conceptualization of the German (but, in
> principle, every) nation as a *human body* that had to be
> *shielded from disease* (or, in case of an outbreak, *cured*).
> Jewish people, who were conceptually condensed into the
> super-category of "the Jew" and viewed as an *illness-
> spreading parasite*, represented the danger of *disease*. Deliv-
> erance from this threat to the nation's *life* would come from
> Hitler and his party as the only *competent healers* who were
> willing to fight the *illness*.[33]

Hitler repeatedly calls "the Jew" a germ, germ carrier, or agent of disease, a decomposing agent, fungus, or maggot. In their capacity as germ-
carriers, Jews are equated with vermin or, more specifically, rats and the
source of an epidemic or pestilence comparable to syphilis. The Jewish
disease is most often presented as a sort of blood-poisoning. "In the
most basic version," Musolff observes, "Hitler likens the Jew to a viper
or adder . . . whose bite directly introduces venom into the bloodstream
of the victim." In their capacity as blood poisoners, Hitler also refers to
Jews as bloodsuckers, leeches, and poisonous parasites (Hitler and Goebbels also characterized Jewry as a "ferment of decomposition," which is
a misappropriation of a phrase from the work of Nobel laureate Theodor Mommsen, who used it to refer to the *positive* contributions that
Jews made to European civilization).[34] Just like the nineteenth-century
polygenecists, the National Socialists clung to the idea that human
races are distinct species. Hitler introduces this theme in *Mein Kampf*
with a juvenile homily about the birds and the bees. "Every animal
mates only with a representative of the same species," he remarks. "The
titmouse seeks the titmouse, the finch the finch, the stork the stork, the

field mouse the field mouse, the common mouse the common mouse, the wolf the wolf, etc."

Have you got wind of where this is heading? Hitler continues:

> The consequence of this purity of race, generally valid in Nature, is not only the sharp delimitation of the races from others, but also their uniform character in themselves. A fox is always a fox, a goose a goose, a tiger a tiger, etc.[35]

The two races are irrevocably distinct, and their mixture is an affront to nature, leading to "a slowly but surely progressing sickness" and producing "monstrosities half-way between man and ape." From a scientific perspective, this is gibberish. Races aren't species, and descriptive biological laws, which cannot be violated, are nothing like prescriptive social regulations, which can. In conflating these categories, Hitler was trying to establish that Jews and Aryans are radically different *kinds* of beings, and to underwrite his genocidal policies not only by nature, but also by God, who created the laws of nature. Racial mixing, he claims, is a "sin against the will of the eternal Creator."[36]

At the same time that the National Socialist project was gathering momentum, the Soviet state was visiting mass murder on its own citizens. The targets were so-called Kulaks (the word literally means "fists," meant as an abbreviation for "tight-fisted"), relatively affluent independent farmers who had been identified by the Soviet regime as class enemies of the communist state. At least nine million were killed. The Soviet propaganda machine ground out political posters that presented them as creatures such as snakes, spiders, and vermin.[37]

Russian dissident journalist Vasily Grossman portrayed the fate of the Kulaks in his mordant novel, *Forever Flowing*. In it, Anna Sergeyevna, a former state-employed killer, recounts:

> They would threaten people with guns, as if they were under a spell, calling small children "kulak bastards," screaming "bloodsuckers!" . . . They had sold themselves on the idea that the so-called "kulaks" were pariahs, untouchables,

vermin. They would not sit down at a "parasite's table"; the "kulak" child was loathsome, the young "kulak" girl was lower than a louse. They looked upon the so-called "kulaks" as cattle, swine, loathsome, repulsive: they had no souls; they stank; they all had venereal diseases . . . they were not human beings.[38]

And in another passage, a former Stalinist killer recalls:

What I said to myself at the time was "They are not human beings, they are kulaks". . . . In order to massacre them it was necessary to proclaim that kulaks are not human beings. Just as the Germans proclaimed that Jews are not human beings. Lenin and Stalin proclaim, kulaks are not human beings.[39]

Next is the Cambodian genocide of the 1970s, engineered and implemented by Pol Pot's Khmer Rouge (Communist Party of Kampuchea). The party seized power in the spring of 1975, after five years of bloody civil war. Once in control, they instituted a massive program of ethnic cleansing that wiped out around a fifth of the population of Cambodia.

The Khmer Rouge instituted sweeping reforms, modeled on Mao Zedong's "Great Leap Forward." Similarly, their "purification" of Cambodia resonated with Mao's Cultural Revolution.[40] The Great Leap Forward was a disastrous attempt to collectivize and modernize Chinese agriculture and industry that resulted in the death of up to 43 million people, mainly from starvation. Its failure led to Mao's resignation as chairman of the People's Republic of China, and a swing to the right in the Chinese communist party. Then, beginning in 1966, Mao mobilized the populace to purge the party of these "bourgeois" elements in a mass movement called the Cultural Revolution. Those suspected of being class enemies were described as undesirable animals: cow ghosts and snake spirits (in Chinese folklore, evil supernatural beings that disguise themselves in human form), monsters and demons, parasites and, of course, vermin. The Red Guard—student militias that enforced Mao's program of social reform—used derogatory labels like "pigs," "dogs," and "vampires" for the men and women whom they persecuted, and imprisoned them in detention centers called cowsheds.[41]

> Many party administrators were labeled "capitalist running dogs" and removed from office. The *Red Flag* published an article titled "Clean Up All the Parasites" (Vol. 11, 1966). Soon slogans such as "Down with monsters and demons" and "Sweeping away all monsters and demons" were everywhere: on wall posters, in official newspapers, in the Red Guards' leaflets and in the rallying cries.[42]

Chemistry professor Li-Ping Luo, who grew up during this period, recounts how dehumanizing talk was intertwined with acts of violence:

> It was still late August when one day the door to the Spanish House was flung open, and the two spinsters were forced to kneel in front of it. Dragged from her wheelchair, the old mother was forced to join them, but she was so feeble that she collapsed into a heap on the terrace. The mob yelled and screamed, banging their fists on every available surface. The old women were called "bloodsucking leeches," "maggots," and "intestinal parasites."[43]

Similar to Mao's cultural revolutionaries, the Khmer Rouge portrayed those regarded as internal enemies as worms, germs, termites, and weevils, infecting and undermining the new political order. Pol Pot commanded his army to cleanse Cambodia of ethnic Vietnamese, enjoining them to "kill the enemy at will, and the contemptible Vietnamese will surely shriek like monkeys screeching all over the forest." This was to be a nation for the Khmer (the majority ethnic group in Cambodia), purged of all foreign influences. Thus began a gigantic, and tragic, experiment in social engineering in which religion, public services, industry, banking, and even currency were abolished. There was to be no private property, and formerly urban populations were forced to become agricultural laborers.[44]

Yale University historian Ben Kiernan, an expert on the Cambodian genocide, explains that ethnic Khmer who were suspected of being unsympathetic to the party's aims were accused of merely *appearing* to be Khmer, of having "Khmer bodies with Vietnamese Minds."[45] Like the Armenian genocide and the Holocaust, the Cambodian genocide was conceived as a cleansing of diseased elements, the extirpation of a lethal infection ("what is infected must be cut out" was a popular party slogan), and the elimination of carriers of germs. As Pol Pot put it, "There is a sickness in the Party. . . . If we wait any longer, the microbes can do real damage." Men and women were executed because the "pro-Vietnamese virus" infected them.[46] Kiernan adds that:

> While making no claims of genetic racism or "scientific" precision, the CPK leadership employed biological metaphors that suggested the threat of contamination. It referred to enemies as "diseased elements," "microbes," "pests buried within," and traitors "boring in," just as the Nazis had talked of Jews as "vermin" and "lice." Pol Pot considered the CPK's revolution the only "clean" one in history, just as the Nazis had "cleaned" areas of Jews; and his regime, equally obsessed with "purity," launched its most extensive massacres . . . with a call to "purify . . . the masses of the people."

And after a bloodbath in the Eastern provinces, in which as many as a quarter of a million Khmer were wiped out, Pol Pot announced, "The party is clean. The soldiers are clean. Cleanliness is the foundation."[47]

Those who were not killed were harshly treated. "We were being treated worse than cattle," reported one survivor, "the victims of methodical, institutionalized contempt . . . no longer human beings." Another reported being told that his mother, who had recently died, was less valuable than a cow: "[Cows] help us a lot and do not eat rice. They are much better than you pigs," he was told.[48] Agents of Pol Pot's socialist paradise who tortured and killed approximately seventeen thousand detainees in Tuol Sleng prison in Phnom Penh (also known as "S-21") considered their victims not as human beings but as, in the words of one survivor, "less than garbage." (As another ex-prisoner put it, "We weren't quite people. We were lower forms of life. Killing us was like swatting flies, a way to get rid of undesirables.") Writing about this episode, historian David Chandler points out, "that dehumanizing the victims made them easy to kill."[49] They could even be eaten. One witness described how, as a child, he had watched as a man was killed with an axe, and then "the cadre . . . opened up the man's chest, he took out the liver."

> One man exclaimed, "One man's liver is another man's food." Then a second man quickly placed the liver on an old stump where he sliced it horizontally and fried it in a pan with pig grease. . . . When the liver was cooked, the cadre leader took out two bottles of rice-distilled whisky, which they drank cheerfully.[50]

Almost twenty years later, in 1994, what is perhaps the most notorious genocide of the late twentieth century exploded in Rwanda. It was the culmination of long-standing tensions between the two largest ethnic groups in the country: the Tutsi and the Hutu. Traditionally, the Tutsi, who were a pastoral people, constituted the ruling class, while the agriculturally based Hutu had a lower position on the social hierarchy. When the Belgians colonized Rwanda in the late nineteenth century, they reinforced the existing social hierarchy by regarding the Tutsis as superior to their Hutu countrymen and providing them with greater social and economic opportunities. Once the Tutsi monarchy was overthrown by a Hutu uprising in 1959, ethnic antagonisms remained. In 1963–64, Hutus killed around 10,000 Rwandan Tutsis, and between 1965 and 1991 Tutsis slaughtered around 150,000 Hutus in a series of incidents in neighboring Burundi. By the early 1990s, tensions were coming close to the boiling point. It was at this time that plans were drawn up to exterminate all the Tutsis in Rwanda.

The genocide was foreshadowed in popular media, principally the magazine *Kangura* ("Awaken!"). A year before the genocide erupted, *Kangura* published a notorious article describing the Tutsi as vile, subhuman creatures.

> We began by saying that a cockroach cannot give birth to a butterfly. It is true. A cockroach gives birth to another cockroach. . . . The history of Rwanda shows us clearly that a Tutsi stays always exactly the same, that he has never changed. The malice, the evil are just as we knew them in the history of our country. We are not wrong in saying that a cockroach gives birth to another cockroach. Who could tell the difference between the *inyenzi* who attacked in October 1990 and those of the 1960s? They are all linked . . . their evilness is the same. The unspeakable crimes of the *inyenzi* of today . . . recall those of their elders: killing, pillaging, raping girls and women, etc.[51]

Then, when Hutu president Juvénal Habyarimana's plane was shot down in April 1994—apparently at the hands of radical Hutus, but blamed on Tutsi assassins—all hell broke loose. There were calls to exterminate the Tutsis, and their Hutu sympathizers, as government-supported civilian militias began what was called a "big clean-up."[52] In the space of three months, around 800,000 Tutsis and moderate Hutus were shot, burned, hacked, and bludgeoned to death by marauding mobs.

Dehumanization played an unmistakable role in these events. As the passage I quoted from *Kangura* indicates, Tutsis were called *inyenzi*—cockroaches—in state-sponsored propaganda. Rakiya Omaar, director of the human rights organization African Rights, affirms, "In Rwanda they referred to Tutsis as cockroaches. They were not human beings. . . . [They said,] 'Don't worry, you're not killing humans like you. You are killing some vermin that belongs under your shoe. You're killing cockroaches," which is why a secret military operation against the Tutsi had the code name "operation insecticide."[53] Tutsi were also called rats, vermin, disease, snakes, and sometimes weeds.

Tutsis also dehumanized Hutus, depicting them as monkeys and gorillas, as vicious, flesh-eating monsters, and collectively as a hyena (in Rwanda, hyenas are regarded as extremely filthy and disgusting, as well as being very dangerous).[54]

Slogans like "Tutsis caused problems and must be exterminated with their eggs," "If you cannot catch the louse, you kill its eggs," and "If you set out to kill a rat, you must kill the pregnant rat" were invoked to justify the murder of women and children.[55] Esperance Nyirarugira, a woman who was raped during the genocide but managed to escape with her life (and whose entire family was hacked to pieces) is quite explicit about this. "Based on what I saw," she told Daniel Goldhagen, "Hutu thought of Tutsi as animals. They did not have the value of a human being. . . . You could pass some people and they shout at you saying, 'Look at that cockroach,' 'Look at that snake.'"[56] The killers confirm this. Elie Ngarambe, who took part in the bloodbath, informed Goldhagen that his comrades "did not know that the [Tutsi] were human beings, because if they had thought about that they wouldn't have killed them. Let me include myself as someone who accepted it; I wouldn't have accepted that they [the Tutsi] are human beings."[57]

Finally, I want to address the role of dehumanization in the genocide that has dominated the first decade of the present century: the mass killing in Darfur, Sudan. The historical background to the Darfur genocide is quite complex. It was the upshot of ethnic conflict between a powerful Arab minority (including the Arab-dominated government in Khartoum) and other ethnic groups collectively called *zurug* ("dark" or "black"). These people are physically distinguished from Arabs not by their skin color, but rather by facial features such as the shape of lips and nose, and, importantly, by their ethnic background.[58]

Darfur has a history of ethnic conflict. As late as the nineteenth century, it was a source of human chattel for the Arab slave trade. More recently, Libya's Muammar el-Qadaffi supported the Arab-dominated Sudanese government in the hope of creating an "Arab belt" across sub-Saharan Africa, while the Arab and Islamic Union, a group that successfully campaigned for the election of Sudan's prime minister in 1986, had an explicit agenda of subordinating the *zurug* to Arab rule and claimed that the "Arab race" had introduced civilization to the region. The rising tide of Arab supremacism led to a Darfurian rebellion in 2003, which the Sudanese government tried to put down with the help of Janjaweed militias. The militias brutally and indiscriminately killed, maimed, and raped civilians, which brought about the death of as many as 400,000 people and the displacement of millions more.[59]

The mayhem in Darfur hasn't been accompanied by written or visual propaganda. There are no films like *The Eternal Jew* or magazines like *Kangura*. But the influence of dehumanization is evidenced by victims' accounts of what the Janjaweed say to them during the attacks.

> "Dog, son of dogs, and we came to kill you and your kids."
> "Kill the black donkeys! Kill the black dogs! Kill the black monkeys!"
> "You blacks are not human. We can do anything we want to you."
> "We kill our cows when they have black calves. We will kill you too."
> "You make this area dirty; we are here to clean the area."
> "You blacks are like monkeys. You are not human."[60]

And so it goes on. As I write these words, there is a fragile peace in Darfur—a peace that might at any point give way to renewed violence. But even if the peace turns out to be lasting, you can be certain that the first major genocide of the twenty-first century will not be its last.

THE SUBHUMAN

> Art thou a man? Thy form cries out thou art . . . thy wild
> acts denote the unreasonable fury of a beast.
> —WILLIAM SHAKESPEARE, *ROMEO AND JULIET*[61]

So far, in this book, I have kept theory to a minimum. My aim has been to convince you that dehumanization is a real and significant phenomenon, and that it is worthwhile taking it seriously. Hopefully, I have succeeded in this task, and it's now time to shift gears. I'll continue to describe examples of dehumanization in the rest of the book, but my descriptions will be more closely tied to an examination of the underlying processes. There will be a little less narrative, and a lot more analysis.

I want to start by considering an exceptionally explicit example. In 1942, a lavishly illustrated magazine hit the German newsstands. Entitled *The Subhuman*, edited by Himmler, and published under the imprimatur of the SS, its purpose was to educate the German public about the alarming threat posed by "Mulattos and Finn-Asian barbarians, Gypsies and black skin savages . . . headed by . . . the eternal Jew," creatures that are "beasts in human form."

The text begins with a quotation from Himmler himself, dated 1935, and this sets the tone for the pages to follow. "As long as there have been men on the earth," he expounds, "the struggle between man and the subhuman will be the historic rule; the Jewish-led struggle against mankind, as far back as we can look, is part of the natural course of life on our planet. One can be convinced with full certainty that this struggle for life and death is just as much a law of nature as is the struggle of an infection to corrupt a healthy body."[62] The text goes on to proclaim:

> Just as the night rises against the day, the light and dark are in eternal conflict. So too, is the subhuman the greatest enemy of the dominant species on earth, mankind. The subhuman is a biological creature, crafted by nature, which has hands, legs, eyes, and mouth, even the semblance of a brain. Nevertheless, this terrible creature is only a partial human being. . . . Not all of those who appear human are in fact so.

> Although it has features similar to a human, the subhuman is lower on the spiritual and psychological scale than any animal. Inside of this creature lie wild and unrestrained passions: an incessant need to destroy, filled with the most primitive desires, chaos and coldhearted villainy.

There are fifty more pages of this stuff, accompanied by photographs of sinister-looking Jews and Russians, filthy hovels, desecrated churches, and piles of corpses (to illustrate the Jewish proclivity for committing mass atrocities). Illustrations of Jewish degradation are juxtaposed with pictures of their wholesome Aryan counterparts. The illustrations of Jews have captions like "subhuman horde," "Jews as subhuman leaders," "the mud huts of the subhuman," and "wanton murder by the subhumans."

This should all seem familiar. Looking back on the examples of dehumanization presented in this book—the genocides described in the present chapter, the oppression and enslavement of sub-Saharan Africans, the extermination of the indigenous peoples of the New World, and all the rest—it's striking that, although the details vary from culture to culture and epoch to epoch, the dehumanizing imagination consistently produces astonishingly similar results. Not only do the images of the dehumanized people resemble one another to a remarkable degree, but the general *form* of dehumanization—the broad assumptions that inform it—are also strikingly similar. This unity in diversity suggests that there is something that all of these examples have in common, something that undercuts their cultural and historical variety. That "something," I will argue, has to do with the machinery of the human mind and the psychological legacy of the evolutionary trajectory of our species.

In the next few chapters, I will be examining these matters in detail. For the moment, I will summarize the fundamental characteristics that all the examples of dehumanization share, illustrating each of them with examples from the Nazi era, and particularly *The Subhuman*, as well as occasionally from other socio-historical contexts.

QUASI-HUMAN KINDS

Dehumanization applies to whole groups, rather than particular individuals: *all* barbarians, *all* Native Americans, *all* Armenians, *all* Blacks, *all* Tutsi. In *The Subhuman* these groups are listed as Mulattos, Finn-Asians (Slavs), Gypsies, Blacks, and, above all, Jews. The collective character of dehumanization is exemplified by some of Himmler's remarks in the Poznan speech. Himmler ridiculed those party members who wanted to make exceptions to the policy that all Jews were to be exterminated.

> This is one of the things that is easily said: "The Jewish people are going to be exterminated," that's what every party member says. "Sure, it's in our program, elimination of the Jews, extermination—it'll be done." And then they all come along, the 80 million worthy Germans, and each one has his one decent Jew. Of course, the others are swine, but this one, he is a first-rate Jew.[63]

In a recording of the speech, you can hear laughter in the audience after the last sentence. For Himmler, as for other committed Nazis, anyone holding up an example of a "first-rate Jew" was missing the point of the Nazi race theory. The Jews were to be eliminated *because they were Jews*. The characteristics of individual Jews, however agreeable, were strictly irrelevant to the exterminationist project. The Jews were a kind of being, but not a human kind of being. Like all dehumanized populations, they were pictured as a *quasi-human* kind—a sort of nonhuman creature possessing superficially humanlike attributes, a monstrous mixture of outwardly human and inwardly nonhuman elements.

APPEARANCE AND REALITY

The very idea that there can be entities with superficially human characteristics, but which are inwardly subhuman, depends on a prior distinction between appearance and reality. Dehumanization would be impossible if we could not make the distinction between how things seem and how they really are. The discourse of dehumanization typically involves the idea that members of a certain population are not what they appear to be. This is usually implicit, but sometimes it is quite overt. In *The Subhuman*, Jews are portrayed as "beasts in human form," even though they have "features similar to a human." They may look like humans and act like them, but inside—where it matters—they are not at all like humans. When one group of people dehumanizes another, they imagine that, although the latter may look human (in the idiom of *The Subhuman*, they have "hands, legs, eyes, and mouth, even the semblance of a brain"), they lack that inner spark or soul that only humans possess. "Not all of those who appear human are in fact so," *The Subhuman* warns, and "Woe to him who forgets it!"

If subhumans are not really human, then what are they? This brings us to the "sub" in "subhuman," which indicates that they exist at a lower rung of the great chain of being. They are lesser biological entities—typically, dangerous predators, poisonous animals, carriers of disease or disease organisms, parasites, traditionally filthy animals, or bodily products, especially feces. In *The Subhuman*, Jews are first described as a vector of infection. The Nazi diatribe goes on to describe Jews as nocturnal animals living in fetid environments.

> The subhuman is united with his peers. Like beasts among beasts, never knowing peace or calm. The subhuman thrives in chaos and darkness, he is frightened by the light. These subhuman creatures dwell in the cesspools, and swamps, preferring a hell on earth, to the light of the sun. But in these swamps and cesspools the subhuman has found its leader— The Eternal Jew!

The text also hints at their predatory characteristics by gesturing toward their "wild and unrestrained passions" and "incessant need to destroy."[64]

Dehumanization does not always conform to the patterns that I have just described, although they are the most common. Sometimes dehumanized people are thought of as domestic animals (as when the Janjaweed refer to black Darfurians as "donkeys"), or as monkeys, baboons, or apes, as the German and Dutch colonists treated the Herero and Nama. More rarely, dehumanized populations are imagined as prey—animals to be hunted for pleasure (we will encounter examples of this when we look at the role of dehumanization in war). Very occasionally, they are portrayed as invasive or undesirable plants (for example, when Hutus referred to Tutsis as "weeds").

Demoting a population to subhuman status excludes them from the universe of moral obligation. Whatever responsibilities we have toward nonhuman animals, they are not the same as those we have toward members of our own species. So, if human-looking creatures are not really people, then we don't have to treat them as people. They can be used instrumentally, with complete disregard for their human worth— they can be killed, tortured, raped, experimented upon, and even eaten.

THE MYTH OF BLOOD: IMMUTABILITY AND HEREDITY

The children are not the enemy. . . . The enemy is the blood in them.

—SS OFFICER OSKAR GROENING

(UPON BEING ASKED WHY JEWISH CHILDREN

WERE KILLED AT AUSCHWITZ)[65]

Subhumanity is typically thought to be a permanent condition. Subhumans can't become humans any more than frogs can become princes. As the text of *The Subhuman* puts it, "The subhuman will always be a subhuman."

There is an important exception to this principle, though. Some religions, notably Christianity, claim to be able to transmute subhumans into human beings, just as, in the Roman Catholic mass, wafers and wine are miraculously transmuted into the body and blood of Christ. But in these cases there is often a tension between the religious notion of redemption through divine grace, and the more basic conviction that subhumanity is immutable. When Jews living in Spain during the fifteenth and sixteenth centuries were forced to convert to Christianity on pain of expulsion or execution, even those who chose to embrace Christianity, at least nominally, were not granted full equality. The so-called "new Christians" were constrained by discriminatory laws called *limpieza de sangre* ("purity of blood") statutes. Jews who converted to

Christianity, and even Christians with Jewish ancestors, remained *marranos* (swine).

Belief in the unchangeability of the subhuman condition is linked to another assumption. Subhumanity is thought to be passed down by parents to their offspring. Even innocuous infants carry the dangerous, subhuman essence within. Typically, the subhuman essence (as well as the human essence) is imagined to be carried in the blood. In this framework, it is vital to prevent human blood from being polluted by subhuman blood. German Nazis (as well as fifteenth-century Spaniards, North American racists, and others) were preoccupied with heredity and the purity of blood to the point of obsession. Even young children were required to memorize poems extolling the purity of their blood, such as this ditty, which was a favorite of Walter Gross, head of the Nazi Party's Racial Policy Office.

> *Keep your blood pure,*
> *it is not yours alone,*
> *it comes from far away,*
> *it flows into the distance*
> *laden with thousands of ancestors*
> *and it holds the entire future!*
> *It is your eternal life.*[66]

The mystical notion of blood-borne subhumanity was the basis for laws against miscegenation (Hitler compared ethnically mixed marriages to "unions between ape and human").[67] Racial mixture "taints" the purity of human blood with a subhuman contaminant. (*The Subhuman* states that Aryan nations that are "tainted by the mixing of blood" are thereby destroyed.)

The notion of blood purity had an obvious bearing on one of the fundamental problems confronting the Nazi program of racial hygiene. Jews often look just like Aryans. Given this, it was vital for the Nazis to hit upon some way to reliably distinguish subhumans from humans. Duke University historian Claudia Koonz describes how, during the early 1930s, German biologists launched a research program with the aim of discovering "the distinctive traits in Jewish blood."

Writing in a popular magazine published by Gross' Office of Racial Politics, a biologist exulted in 1934, "Think what it might mean if we could identify non-Aryans in the test-tube! Then neither deception, nor baptism, nor name change, nor citizenship, and not even nasal surgery could help! . . . One cannot change one's blood." Despite significant funding and considerable publicity, success eluded them. Even a task-force of the Nazi medical association led by Gerhard Wagner admitted failure. No blood-type, odor, foot- or fingerprint pattern, skull size, earlobe or nose shape, or any other physiological marker of Jewishness withstood scrutiny.[68]

The abject failure of scientific efforts to pin down exactly what it is that makes Jews Jews caused Hitler to reorient his racial policy. Koonz remarks "From that point on . . . the burden of proof shifted away from the natural sciences to the social sciences and the humanities."[69] The new race experts were literary and legal scholars, linguists, historians, geographers, and anthropologists, rather than physicians and biologists.

Hitler explicitly denied that race can be biologically defined. He set this out clearly in a letter dictated in February 1945 to his private secretary, Martin Bormann.

We use the term Jewish race merely for reasons of linguistic convenience, for in the real sense of the word, and from a genetic point of view there is no Jewish race. Present circumstances force upon us this characterization of the group of common race and intellect, to which all the Jews of the world profess their loyalty, regardless of the nationality identified in the passport of each individual. This group of persons we designate as the Jewish race. . . . The Jewish race is above all a community of the spirit. . . . Spiritual race is of a more solid and more durable kind than natural race. Wherever he goes, the Jew remains a Jew . . . presenting sad proof of the superiority of the "spirit" over the flesh.[70]

Having given up on biology, Hitler embraced a vague, nonbiological notion of heredity. However, a moment's reflection shows that this strategy doesn't solve the problem at all. Instead, it creates what philosophers call an "infinite regress." Suppose that a Jew is anyone with Jewish parents. Then you're stuck with the problem of how to determine if that person's parents are Jewish. To do this, you need to figure out if *their* parents were Jewish, and so on, ad infinitum. To make matters worse, many Germans had Jewish ancestors, so the Nazis needed to quantify just how much of a Jewish lineage makes one a Jew. Having concurred that "no satisfying biological solution [to this problem] exists," Hitler decided to define as "Jewish" anyone with three or four Jewish grandparents. This was a pragmatic and obviously arbitrary solution. "Right now, we're not discussing utopias," he remarked, "but we're looking at the day-to-day reality and the political situation in the eye."[71]

Having identified some of the main features of dehumanization, we are finally in a position to explore why it has such a powerful grip on the human psyche. In the next chapter, I will show that this terrifying, destructive, and tragic dimension of human existence stems from perfectly ordinary psychological processes and inclinations. Ironically, capacity to dehumanize members of our own species turns out to be rooted in our distinctively human psychology.

6

RACE

———

Listening . . . a slow horror would drain the blood from
my throat and I would think, He is saying these things not
caring what they hear, as though they were not human!
And then would come the tearing dialogue. Well, are
they? are they? of course they are! they're just like me. but
are you sure? Of course! then why are they not in our
church? why are they not in our school? Why can't we
keep playing together? what is wrong what is wrong?
— LILLIAN SMITH, *KILLERS OF THE DREAM*[1]

WITH THESE WORDS, civil rights activist Lillian Smith described her
memory of standing in a crowd in a small Georgia town, at the dawn of
the twentieth century, listening to a segregationist politician harangue
his audience "telling us lies about skin color . . . lies made of their own
fantasies, of their secret deviations—forcing decayed pieces of theirs and
the region's obscenities into the minds of the young and leaving them
there to fester."[2]

Smith's account highlights the way that dehumanization is sometimes overtly intertwined with racism. In this chapter, I am going to
argue that dehumanization probably is, despite appearances to the contrary, *always* bound up with racism. In fact, the concept of race is the
place where the psychological, cultural, and ultimately biological dimensions of dehumanization all converge. Attending to it not only

gives us a clearer conception of what dehumanization is all about. It also throws new light on notions of race and racism.

THE PUZZLE OF RACE

To most of us, the reality of race seems to be so self-evident as to be unquestionable. In fact, the idea of race is reinforced day in and day out. Census forms ask us to check the box corresponding to our race (sadly, there is no box available for those to check who identify themselves as members of the *human* race). Universities offer courses in Black Studies (and, increasingly, in Whiteness Studies). Antidiscrimination laws forbid employers from excluding job candidates on the basis of their race. We are surrounded by practices, attitudes, and institutions that *presuppose* the reality of race.

To most people, questioning the concept of race is likely to sound as crazy as questioning the claim that the Earth revolves around the sun. It seems self-evidently true that there are races. Just look around! However, science presents us with quite a different picture of the mosaic of human variation. Although there are people with striking physical similarities, conventional racial categories of the sort that loom so large in nonscientific discourse—categories of the sort represented by the boxes ticked on census forms and job applications—don't have scientific justification. This was pointed out as early as 1935 by biologist Julian Huxley, anthropologist Alfred Court Haddon, and zoologist/sociologist Alexander Carr-Saunders in their book *We Europeans*, which they wrote in response to the rise of fascism in Europe. These authors made it clear that, from a biological standpoint, the racial classifications that most people take for granted are sheer fiction, on a par with elves, trolls, and fairies, and work in the biological sciences since 1935—in particular, dramatic progress in molecular biology—has amply confirmed their assessment.

For the standard racial categories to make biological sense, racial boundaries should be circumscribed by genetic commonalities. In other words, any two members of a "race" should have more genetically in common than any two people belonging to different races. But this isn't so. It's quite possible that I (a light-hued, blue-eyed person of mainly European descent) am more genetically similar to my wife (a dark-hued, brown-eyed person of mainly African descent) than I am to someone whose skin and eye color more closely match my own. As Huxley and his coauthors put the point over seventy years ago, "With a species in which intercrossing of divergent types is so prevalent as our own, no simple system of classification can ever be devised to represent the diversity of the situation." Biologists can and do sometimes talk about "races" as local breeding populations, but this has nothing to do with the typological notions that suffuse popular culture.[3]

Nevertheless, the race concept not only *seems* to make sense, it is extraordinarily compelling. After all, people are physically different from one another, in a number of obvious ways, and these differences allow us to group them into categories based on similarity. These are bodily differences passed on biologically from parents to children, and are distinct from acquired cultural differences such as language, religion, and forms of dress. Aren't these differences the basis for racial membership? People belong together, racially, because they resemble one another. What could be more obvious?

But look closer, and the seemingly commonsensical idea that obvious similarities are the basis for the race concept is revealed as chimerical. Think for a moment about similarity. Any two people are similar in some respects and dissimilar in others, and *many* of these similarities and dissimilarities are inherited from parents and passed on to offspring. This is broadly true of conventional racial markers like skin color and hair texture. But why single out certain traits as the basis for racial classification, while ignoring others? Couldn't we equally well use eye color, dividing humanity into blue-eyed, brown-eyed, and green-eyed "races"? How about height or earlobe size, or combinations of all traits? Why privilege nose shape over eye shape when distinguishing blacks from whites, while privileging eye shape over nose shape when distinguishing whites from East Asians? Why carve up the human world in just these ways, when there are so many similarities and dissimilarities to choose from?

WHAT MADE JANIE BLACK?

The fact that race isn't just a matter of how people look is actually quite a commonplace idea. For much of U.S. history, conventional ideas about race produced consequences that were at once tragic and absurd. Writer and social critic Lillian Smith gave us an unforgettable example of one such episode in *Killers of the Dream,* her deeply moving memoir of growing up in Georgia during the early decades of the twentieth century. "A little white girl was found in the colored section of our town," she recalled, "living with a Negro family in a broken-down shack." Members of a white women's club concluded that the child (who is described as "very white indeed"—that is, *visually* white) must have been kidnapped, and persuaded the local marshal to take her into custody. The little girl, named Janie, came to live with the Smiths, where she and Lillian quickly became close friends, until a surprising phone call from an African-American orphanage revealed that Janie was, in "fact," black. This discovery changed everything. In an instant, Janie was transformed into a black child who had to be extruded from the family circle.

> In a little while my mother called my sister and me into her bedroom and told us that in the morning Janie would return to Colored Town. . . . And then I found it possible to say, "Why is she leaving? She likes us, she hardly knows them. She told me she had been with them only a month."
> "Because," Mother said gently, "Janie is a little colored girl."
> "But she's white!"
> "We were mistaken. She is colored."
> "But she looks—"
> "She is colored. Please don't argue!"
> "What does it mean?" I whispered.
> "It means," Mother said slowly, "that she has to live in Colored Town with colored people."

In befriending Janie, Lillian transgressed an ironclad social taboo, the incoherence of which made it no less absolute, and she was consequently wracked with guilt.

> I was white. She was colored. We must not be together. It was bad to be together. Though you ate with your nurse when you were little, it was bad to eat with any colored person after that. It was bad just as other things were bad that your mother had told you. It was bad that she was to sleep in the room with me that night. It was bad. . . . [4]

What, if anything, made Janie black?

There are two possibilities here. One is to say that Janie wasn't black, or white, or anything else. According to this skeptical approach, racial categories are both false and dangerous, and should be expunged from our vocabulary. The other option is to say that Janie was indeed black, but her blackness was a social fact rather than a biological one. According to this view, which is known as *social constructionism*, races are real, but they are artifacts of social classification. Janie was black just because she was classified as black. Nowadays, social constructionism is the prevailing orthodoxy in the study of race.[5]

Although skeptics and constructionists are fundamentally at odds about the concept of race, this shouldn't obscure the fact that they also share a lot of common ground. They both agree that racial classification is a real and powerful force, and hold that racial categories don't drop out of the blue, but are ideological constructions that depend on particular cultural and historical circumstances. The difference between the two views turns on the question of whether racial categories should be given any credence.[6]

The fact that racial classifications are ideological explains how and why they change over time. The history of racial taxonomy in the United States provides some revealing examples. Today, there are six officially recognized races: White, American Indian and Alaska Native, Asian, Black or African American, Native Hawaiian, and Other Pacific Islander (although being Hispanic is considered an "ethnicity" rather than a race in government documents, we will shortly see that this is a distinction without a difference). The racial spectrum was very different early in the twentieth century. At that time, Jews, Irish, Slavs, Italians, and a ragbag of others were classified as separate races, and these races were thought to pose a serious threat to "white" hegemony. MIT president Francis A. Walker, writing in 1896, described these groups as "beaten races; representing the worst failures in the struggle for existence" and emphasized that they "have none of the ideas and aptitudes that fit men to take up readily and easily the problem of self-care and self-government." Walker was just one voice in a panicky nativist chorus calling for restrictions on *European* immigration, lest racially inferior specimens of humanity overrun the country.[7]

Why have our notions of race changed so dramatically over the last hundred years or so? It's not because we've discovered any new biological facts about which groups are *really* races and which ones aren't. No white-coated geneticist has emerged from the lab to proclaim, with the authority of science behind him, that Italians are white and Nigerians are black. What's occurred is a conceptual shift, caused by sociopolitical changes.

Changes in the racial classification of Native Americans between the sixteenth and nineteenth centuries provide an especially compelling example of the ideological function of race. During the sixteenth and early seventeenth centuries, when English settlers were on relatively good terms with Native Americans, they described the latter as white (their "olive" or "tawny" appearance was attributed to exposure to the sun or to ointments that they applied to their bodies). However, as tensions mounted, and episodes of violence became more frequent and sustained, settlers began to describe Indians as "copper colored" or "red" rather than fair skinned. At the same time, colonists amplified Native Americans' alterity in other ways. New legislation placed Indians in the same category as blacks and mulattos—that is, as racially "other"—and forbade intermarriage of settlers and Native Americans, defining it as a form of miscegenation. By the early eighteenth century, Indians had undergone an astonishing metamorphosis. They were no longer white, but red. According to historian A. T. Vaughan, by the beginning of the nineteenth century, "the stereotypical color

carried a host of unfavorable connotations that prevented the Indians'
full assimilation into the Anglo-American community and simultaneously precluded their acceptance as a separate and equal people."

Although a few dissenters resisted the prevailing color
taxonomy and its correlative racial policies, the surviving
literature, both factual and fictional, shows that the Indian
was no longer considered a member of the same race; he
remained forever distinct in color and character. Even relatively sympathetic spokesmen now believed the Indians to
be permanently different.[8]

BEYOND IDEOLOGY

I suppose it is a truth too well attested to you, to need a
proof here that we are a race of beings, who . . . have long
been considered rather as brutish than human, and scarcely
capable of mental endowments.
              —BENJAMIN BANNEKER TO THOMAS JEFFERSON,
                                        AUGUST 19, 1791[9]

Social constructionism accounts for the fluidity and historical specificity of beliefs about race. But it also has limitations. Its chief shortcoming,
as a comprehensive theory of race, lies in what it doesn't address. Constructionism does a good job explaining the *content* of racial thinking,
but lacks the resources for explaining its distinctive *form*. The practice
of segregating the human species into races reflects a certain style of
thinking. Social constructionism explains why we classify certain groups
as races, but it has nothing to say about why the very concept of race is
so widespread and historically persistent, and it does not address the
question of why it is that the racial concepts deployed by culturally and
historically diverse societies have so much in common.[10]
    Some social constructionists claim that the very notion of race is an
ideological invention tied to a particular historical epoch. Some argue
that it originated in the fifteenth century, with the expulsion of Jews

and Muslims from Spain and the imposition of "purity of blood" laws. Others believe that it began in the sixteenth century, as an offshoot of European colonialism. Some locate it in the consolidation of European class relations and the expansion of the transatlantic slave trade during the seventeenth century. Yet others claim that it originated in nineteenth-century biology and anthropology.[11] All of these theorists hold that prior to the comparatively recent "construction" of race, racialism did not exist, but the sheer lack of convergence on a single historical epoch when the notion of race was supposedly "constructed" ought to make us suspicious that they are barking up the wrong tree. I've already presented evidence against the extreme version of the constructionist position in my discussion of slavery, so I won't repeat it here. Suffice it to say that this version of social constructionism confuses historically specific manifestations of racism with the deeper phenomenon that they are all manifestations *of*. It is certainly correct to say that there are many socially constructed concepts of race, but they are all variations on an underlying theme. Social constructions of race are constrained by the psychology of racial thinking—after all, they weren't constructed ex nihilo. To fail to grasp this is to fail to understand the concept of race and the prevalence of racial beliefs.

Over the past twenty years or so, a new cognitive-evolutionary approach to the study of race has emerged.[12] Theorists in this camp accept that racial categories don't have any scientific justification, and allow that social forces fill out the content of racial categories. But they go beyond the social constructionists, arguing that the near-universality of the concept of race suggests that it reflects something about how the human mind works. If this is right, then it has serious social implications. Lawrence A. Hirschfeld, an anthropologist at New York's New School of Social Research who is one of the foremost researchers into the psychology of race, tells us why. "After almost fifteen years of working on the mental representation of race," he comments, "the conclusions that I've come to are in many respects disquieting. . . ."

> Race is not simply a bad idea; it is a deeply rooted bad idea.
> This is not an appealing thought. It implies that race may
> be as firmly grounded in our minds as it is in the politics of

our day. . . . Many prefer to believe that race is an accident
of how we happen to categorize the world. I suppose that
this preference alone accounts for why so many people continue to believe that race is not only a bad idea, but a superficial one—one that could be "set straight" by simply correcting
the misinformation that we receive as children, by extolling
the virtues of our diverse world.

But this is not the case. In fact, Hirschfeld's work, which focuses on
children's concepts of race, demonstrates that "children . . . are more
than aware of diversity; they are driven by an endogenous curiosity to
uncover it."

Children . . . do not believe race to be a superficial quality
of the world. Multicultural curricula aside, few people believe that race is only skin deep. Certainly few three-year-
olds do. They believe that race is an intrinsic, immutable,
and essential aspect of a person's identity. Moreover, they
seem to come to this conclusion on their own. They do not
need to be taught that race is a deep property, they know it
themselves already.[13]

This doesn't mean that the idea of race is innate or inevitable. We
are not condemned to be racialists. However, as Hirschfeld aptly points
out, it does imply that we're all *susceptible* to it. The analogy to disease
is quite a telling one. We are vulnerable to certain diseases because of
our biological design. This explains why it is that certain kinds of microorganisms can gain a foothold inside our bodies if we're exposed to
them. By the same token, our mind design makes us vulnerable to the
racial beliefs that that we are exposed to by our culture.

PLATO'S JOINTS

No wonder St. Patrick drove all the venomous vermin
out of Ireland! Its biped mammals supply that island its

full average share of creatures that crawl and eat dirt and
poison every community that they infest.
— GEORGE TEMPLETON STRONG, IN *GEORGE TEMPLETON*
STRONG'S DIARY OF THE CIVIL WAR 1860–1865[14]

Although racial beliefs can be very diverse, there are features that they
all have in common. Think of these as components of a template that
determines the form of our beliefs about race.

The first component is so obvious that it is easy to take for granted.
Races are conceived as *human kinds* (or, when they are dehumanized,
quasi-human kinds). But they are special sorts of human kinds, for not
every human kind is thought of as a race. Firemen are a human kind.
So are people who root for the Red Sox, or who have more than two
children. But nobody thinks of these groups as races. So, being a human kind is necessary but not sufficient for being a race.

We think of races as *natural human kinds*. I've already visited the notion of natural kinds in Chapter Three. There is a sprawling and often
highly technical philosophical literature on this subject. Much of this
literature is concerned with the question of how scientists and philosophers ought to think of natural kinds. In this book, I am not concerned
with this sort of question. Instead, my project is a descriptive one. I want
to focus on the role that natural kinds *actually* play in our everyday,
pretheoretical thinking. To understand dehumanization, we need to
focus on thinking in terms of natural kinds as a psychological phenomenon, rather than as a scientific or philosophical practice.

With this in mind, let's revisit the concept of natural kinds.

We think of the world as divided into types of things, and we give
these types names. Some of these conceptual divisions are thought to
correspond to the structure of the world, carving nature at its joints, as
Plato famously put it, while others are products of human artifice, gerry-
mandered to suit our needs. The former are natural kinds and the latter
are artificial kinds.[15] You probably take it for granted that apple trees are
a natural kind, but things that cost seventy-nine cents a pound aren't.
Why? Well, one striking difference between the two categories has to
do with what philosophers call their "inferential potentials." Knowing
that something is an apple tree gives you a lot of information about it. If

you know that something is an apple tree, you can infer that it has alternately arranged oval leaves, produces pinkish-white blossoms in the
spring, bears fruit with a certain appearance and taste that ripen in the
fall, grows from seeds, is unlikely to grow more than forty feet tall, and
so on. In contrast, knowing that something costs seventy-nine cents a
pound tells you nothing about it apart from its price. It's an inferentially
anemic category. Similarly, identifying an animal as a porcupine provides a wealth of information about its anatomy, physiology, mating
behavior, life cycle, and diet. Correctly classifying a piece of jewelry—
say, a ring—as gold allows you to predict that it will react to aqua regia
to form chloroauric acid, that it will melt if heated to 1947.52 degrees
Fahrenheit, that it has a tensile strength of 120 megapascals, and so on,
while correctly classifying it as a ring tells you precious little about it.
Of course, you need to have specialized knowledge about apple trees,
porcupines, and gold to make the sorts of inferences that I've described.
But the point is that if you had this knowledge then you could make
those inferences. You could *in principle* make them, even though you may
be unable to in practice. Compare this with things that cost seventy-nine
cents a pound, and things that are rings. There just isn't enough to know
about these things to allow *anyone* to make comparably rich inferences
about them.

What is it that makes these inferences possible? Remember, our focus is on the psychology of natural kinds, so we're concerned with what
people *imagine* it is that makes natural kind concepts so inferentially
potent, rather than what scientists and philosophers say about the issue. As it happens, there's been quite a bit of research into this question over the past few decades, and most of it converges on the same
conclusion. We are strongly inclined to imagine that natural kinds
have a hidden essence, and this essence is supposed to explain observable similarities that hold between all members of the kind. The relationship between natural kinds and essences goes the other way around,
too: if something is believed to have an essence, then it's thought to be
a member of a natural kind. So, people tend to assume that having an
essence is a necessary and sufficient condition for something to be a
member of a natural kind.[16]

Let's use this notion to explore the question of race. At a first approximation, *the concept of a race is the concept of a human natural kind.* The members of a race are imagined to possess a common racial essence, an essence that is unique to them and which makes them the kind of people that they are. This general idea is elegantly captured by a provocative thought experiment devised by Northwestern University philosopher Charles W. Mills. Mills's thought experiment concerns a man named Mr. Oreo, who "cannot even think of passing [as white], being quite dark with clearly black African features and with known black ancestry." But Mr. Oreo "is unhappy with his racial designation, so he fills in 'white' on bureaucratic forms, identifies himself as white, and rejects black culture."[17]

Do Mr. Oreo's attitudes and actions make him white?

Mills points out, correctly in my view, that most people would say that Mr. Oreo is really black. He's just pretending to be white, or has a false belief that he is white, or is confused about his race, but he's really black. As a modest, and admittedly unscientific experiment, I've run this scenario past my students countless times, and on each occasion, they have insisted univocally that Mr. Oreo is black, no matter what he says or believes about himself. If you share this intuition, then you are committed to a certain view of what race consists of. Clearly, if you believe that Mr. Oreo is black you must reject the idea that race is a matter of one's subjective "identity," and you must believe that being a member of one or another race is a *fact* about a person. This might be either a biological fact or a social fact, but it's a fact.

Next, Mills revs the experiment up a notch by introducing the *Schuyler machine.* The Schuyler machine is a fictional contraption that modifies black people's appearance in such a way as to make them indistinguishable from white people (a conceit borrowed from George Schuyler's fiercely satirical novel *Black No More*). Suppose that Mr. Oreo undergoes the Schuyler treatment, and emerges with a stereotypically Caucasian appearance: pale skin, straight blond hair, steely blue eyes, thin lips, an aquiline nose, and narrow nostrils.

Has he become white?

Bear in mind that we're not looking for the *facts* about Mr. Oreo's race. The story is just a way for us to get access to our intuitions. You're probably drawn to some belief or other about whether Mr. Oreo is black or white after the treatment.

Which is it?

It's hard to resist the feeling that even after the Schuyler treatment Mr. Oreo remains a black man. As we've already seen, this conclusion has nothing to do with his subjective sense of identity (remember, he doesn't regard himself as black). It seems natural to think that, after emerging from the Schuyler machine, Mr. Oreo only *looked* white but *hadn't really become white*. At first glance, this conclusion seems puzzling. After all, he's now indistinguishable from a white man. In other contexts we're happy to follow the adage that if something walks like a duck and quacks like a duck then it's a duck. Why not when it comes to race?

To highlight the cogency of the question, Mills invites us to "compare another kind of physical transformation, that of bodily physique and strength." Enter the Schwarzenegger machine.

If a machine were invented (call this the Schwarzenegger machine) that could transform 98-pound weaklings into massively muscled supermen capable of pressing hundreds of pounds without the tedium of special diets and weight training, would we say that the person only *looked* strong but had not really *become* strong? Obviously not. His new body, new physique, new strength are real.[18]

Most people would affirm that the man emerging from the Schuyler machine is still *really* black, but would deny that the muscleman emerging from the Schwarzenegger machine is still *really* weak. This difference demands an explanation. I think that the answer lies in the fact that we tend to think—perhaps in spite of ourselves—that black people constitute a natural kind, whereas weak people don't: we're intuitive essentialists about race, but not about muscles. This dichotomy is suggested by the way that we customarily talk about these characteristics. We say that a person *has* large muscles, but we say that they *are* of a certain race (as the economist and race theorist William Z. Ripley put it in 1899, "Race denotes what man is; all . . . other things denote what he does").[19] A person can gain or lose muscle while remaining the same person, but we tend to think that if they were to change their race, it would amount to their becoming an entirely different person.

ESSENTIAL DIFFERENCES

Mills's thought experiment suggests that we tend to intuitively conceive of a person's race as *necessary* to their identity, whereas body build is thought to be contingent—merely a matter of appearance. Philosophers often describe necessity using the jargon of possible worlds. Possible world talk can sound a bit odd at first, but once you get the hang of it, it becomes a very useful tool for capturing what are called *modal intuitions*—ideas about what is possible, impossible, necessary, or contingent.

Here's how it works. Imagine that there is a world corresponding to every way that the actual world could be. In other words, instead of saying "so-and-so is possible," we say "there is a world at which so-and-so." For example, porcupines might have been pink, instead of their standard grayish hue. This can be described in possible world lingo by the sentence, "There is a world at which porcupines are pink." Now, think of the sentence, "All circles are round." This is what philosophers call a *necessary truth*, which means that it *has* to be true—there's just *no way* for a circle to be square or hexagonal or any shape other than round, because roundness is part of what it is to be a circle. We can use possible world jargon to express this by saying, "Circles are round at every world"—we simply can't conceive of a world where circles aren't round.

Now, let's apply this way of thinking to the notion of race.

If a person's race is part of their essence—if they couldn't be a different race while remaining the same person—we can express this by saying that *a person's race remains the same at every world where that person exists*. Mills comes close to this formulation, stating, "For racial realists, people characterizable by their phenotype in our world, with its peculiar history, as belonging to a particular 'race' will continue to have the same 'racial' intellectual and characterological traits in another world with a radically different history."[20] However, Mills's formulation contains a serious flaw. We've seen from the fictional case of Mr. Oreo—as well as the real-life case of Janie—that phenotypic traits (a biological term for observable, physical traits) are only contingently associated with race. It's possible to be "black" without having dark skin, because skin color isn't an essential feature of "blackness" (in possible world talk, there's at least one world where black people have pale skin). Although a person's race remains the same at every world where they exist, their skin color and other phenotypic traits don't. There are numerous cases of "black" people "passing" as white—and even people being classified as "black" at birth but "white" at death. Lawrence Hirschfeld explains that "historical and experimental research has revealed that visual differences in appearance do not map well onto racial categories."

> The same individual's race would be quite different depending on whether he or she were born in Brazil, the United States, or South Africa, or whether he or she were born in the United States in 2001 or 1901. Within a single system of classification, race is permeable. The historian Linda Gordon . . . documented an intriguing account of the racial transformation of a group of children in the early twentieth century who were not White when they left New York, but who were White when they arrived in Arizona one week later. Hahn, Mulinare, and Teutsch . . . examined race identification of all infants who died before their first birthday in St. Louis between 1983 and 1985. Despite the fact that these were the same infants, they found that significantly more were Black when they were born than when they died.

How is this possible? The likely answer turns on the contrast between presumed essence and actual appearance.

Hahn et al. attributed the inconsistency to differences in the way that race is determined at birth and death: At birth race is identified by parents, at death by a physician. Self-identification is based on genealogy, whereas other-identification is based on appearance.[21]

We've already seen that natural kind categories are supposed to provide inferential leverage. Just knowing that something is a member of a certain natural kind is supposed to provide a key to inferring various other facts about it. This is certainly true of ordinary beliefs about race. People tend to assume that knowing that someone belongs to a certain race opens the door to lots more information about them. This assumption lies at the root of racial profiling.

Here's an example of the sorts of inferences that racial categorization licenses. During the late Middle Ages many European Christians believed that Jewish men menstruated. For example, the thirteenth-century writer Jacques de Vitry wrote in his *History of Jerusalem* that Jews "have become unwarlike and weak even as women, and it is said that they have a flux of blood every month," because ". . . God has smitten them in their hinder parts, and put them in perpetual opprobrium" as punishment for the murder of Christ.[22] The unfortunate condition of these men's "hinder parts" was completed by excruciating menstrual cramps that could only be relieved by drinking the blood of butchered Christian children. So, the myth of Jewish male menstruation was bound up with another widespread Christian belief—the "blood libel" that Jews first torture and then kill Christian children, and use their blood to make matzos for the Passover meal.

This theme of male menstruation appears in Bernard Malamud's 1966 novel *The Fixer*, which is based on the arrest and trial of a Ukrainian Jew named Menachem Mendel Beilis on charges of ritual murder in 1913. The protagonist, a Jewish handyman named Yakov Bok, is arrested on the charge of having ritually slaughtered a Christian child. Bok's persecutors expect him to menstruate while imprisoned and awaiting trial. We can use this fictional account to explore the belief that knowing a person's race gives us additional information about them. Malamud writes:

The days were passing and the Russian officials were waiting impatiently for his menstrual period to begin. Grube-
shov and the army general often consulted the calendar. If it
didn't start soon they threatened to pump blood out of his
penis with a machine they had for that purpose. The machine was a pump made of iron with a red indicator to show
how much blood was being drained out. The danger of it
was that it didn't always work right and sometimes sucked
every drop of blood out of the body. It was used exclusively
on Jews; only their penises fitted it.[23]

Suppose that the authorities had administered the pump, and caused
Bok's penis to hemorrhage. If they had done this, they wouldn't have
been confronted with any evidence to contradict their peculiar supposi-
tion about Jewish men. But suppose they didn't use the pump, and had
carefully observed Bok, and seen that he didn't menstruate. Would this
have caused them to abandon their belief that Jewish men menstruate?
I doubt it very much; let me explain why.

Essences aren't observable, and their unobserveability is a logical
consequence of their explanatory role (they're supposed to "lie behind"
observable attributes). Essences are supposed to be responsible for the
observable attributes that a thing displays—at least those attributes that
are typical of its kind. For example, the typical attributes of an apple
tree (its size, shape, the configuration of its leaves, its fruit-bearing capacity, and so on), all of which we can see, are imagined to be effects
of its essence, which we can't see.

An apple tree that has all of these features is, so to speak, true to
its kind. But what about an atypical apple tree: one that isn't true to its
kind? A seedling apple tree may never reach full size if it is deprived of
the sunlight, water, and nutrients that it needs to grow. When this occurs, it is natural to think of the tree as having *failed to realize its essential nature*. The stunted plant hasn't stopped being an apple tree, but it's
an anomalous or malformed specimen, rather than a normal or stereotypical one. This pattern of thinking is typical of folk biology. As cognitive scientist Dan Sperber aptly notes, we reason about animals in quite
a different way than we reason about manufactured items like furniture.

We tend to imagine that the typical features of the kind are somehow *contained* in the members of that kind, even if they're not expressed. "If an animal does not actually possess a feature ascribed to it by its definition," he writes, "then it possesses it virtually: not in its appearance but in its nature." Scott Atran, an anthropologist at the National Center for Scientific Research in Paris, adds:

> This is more than a grammatical point: we can say of a tiger born without legs that it didn't ever get *its* legs, but not of a bean-bag chair that it didn't ever get *its*. Sperber further implies that, say, a plucked bird is still thought to have *its* feathers "virtually" just as a coneless pine "virtually" has *its* cones.[24]

Racial thinking follows more or less the same path. This is why Bok's failing to menstruate is simply irrelevant to the belief that Jewish men menstruate. According to the peculiar logic of race, Bok could be seen as a deviant or defective specimen of Jewry who doesn't menstruate but nevertheless has it "in him" to do so.

I chose the example of male menstruation because it is patently absurd. However, we often think about race in much the same way without noticing the weirdness of it—we think that there are "typical" blacks, "typical" Jews, "typical" Asians, and so on. Individuals that depart from the stereotype are considered deviant specimens of their race. A Jewish farmer is (in Europe and the United States, anyway) seen as "less Jewish" than a Jewish accountant—and an African-American philosopher is thought of as "less black" than an African-American athlete.

Because of our psychological makeup, it's all too easy to assume that for any putative natural kind, including racial kinds, certain observable features—those regarded as typical of the kind—are brought into being by the flowering of the kind's essence, and atypical features are imagined to be an effect of some obstruction that prevents the essence from expressing itself in a "pure" form.

Once we come to recognize how all this works, it becomes clear why Janie was considered black even though she looked white, and why Mr.

Oreo is still considered black after the Schuyler treatment. We imagine that in these cases the racial essence was present but failed to manifest. Mr. Oreo concealed his essence by cultivating a misleading appearance, and Janie's essence—her "blackness"—was never fully expressed, but it was assumed that, beneath the surface, they were both really black, and that this was a permanent condition.

There are also folk-theories about how racial essences are carried and transmitted from parents to their offspring. Sometimes, people are uncommitted to one or another theory of racial transmission, but have "an intuitive belief that an essence exists, even if its details have not yet been revealed."[25] But this is by no means always, or even mostly, the case. Yale University psychologists George E. Newman and Frank C. Keil have found that during their earliest years, children assume that the essence of a thing is found in its center. Later on—around the age of ten—they transition to the view that a thing's essence is distributed throughout its body. Newman and Keil inform us that:

> Children, as young as 6 years old, do not seem to be agnostic about the physical nature of essence. However, these younger children, contrary to adults, favor the view that essences are localized to the center of objects—not only for animals, but for minerals as well. Around second grade, children begin to shift away from this Localist view to recognize that for minerals at least, essential features are distributed throughout. By fourth grade, children, like adults, recognize that for both minerals and animals, essential features are distributed—a view which they apply to natural kinds, but not to artifacts.[26]

Although most older children and adults embrace a folk-theory that biological essences are distributed throughout an animal's body, they also retain a residue of the localist theory. The *individual essence* of a thing (that which makes you the individual person that you are, rather than a member of the human *kind*) is typically pictured as a "soul" located in the body—usually in the head just behind the eyes, or in the heart.

The idea that essences are distributed throughout the body flows naturally into the near-universal view that racial essences are carried by bodily fluids. The most popular belief of this sort, which I discussed briefly in Chapter Five, is that they are carried in a person's blood and transmitted down their "bloodline." This theory that a person's racial essence is contained in their blood was invoked by the "purity of blood" laws in fifteenth-century Spain and Germans of the Third Reich, and it's also a feature of numerous traditional belief-systems (for example, the Jívaro of the Andes foothills of Ecuador believe that a living person's true soul is located in their blood, and that bleeding is therefore a loss of soul). Even today, in the developed world, it's not unusual for educated people to assert that they have "African blood" or "Native American blood," that they are a "full-blooded" member of some race or ethnic group, or that certain characteristics are "in their blood."[27]

The notion that race is somehow in the blood has sometimes led to anxieties about the consequences of mixing blood, and not just in the context of antimiscegenation laws. When blood transfusion first became available during World War II, efforts were made to ensure that "black blood" was not given to white servicemen.

> At first the Red Cross announced in November 1941 a policy of excluding black donors, but after the outcry that ensued, it compromised by agreeing to accept blood from black people on the condition that it be kept segregated from the blood of whites. Scientists knew conclusively that this was not necessary, but the surgeons general of the Army and Navy and officials of the Red Cross believed that the program would not work otherwise. Too many white soldiers believed that skin color and "racial traits" could be transmitted through the blood. . . . [28]

And according to Bertrand Russell, Nazi soldiers were terrified of the possibility of receiving transfusions of blood that had been taken from Jews, writing that careful steps were taken to prevent this from occurring.[29]

Another variant on the theory is the idea that racial essences are distributed in breast milk. In many cultures, breast milk is believed to be formed from a woman's blood, so drinking it amounts to sharing her blood, and establishing kinship in a "horizontal" fashion. "In Islam," writes anthropologist Aparna Rao, "the kinship of milk (*rida'a*), like that of blood, restricts marriage between certain persons; by the same token it also functions, as blood does, to broaden bonds between individuals and groups and draw these all into one big family."[30] Rutgers University medical historian Janet Lynne Golden points out that in nineteenth-century America:

> Some believed that children literally drank up their wet nurses' moral and physical imperfections. . . . Mary Terhune narrated the story of a girl said to be "remarkably dissimilar" from other members of her family with "rough skin, corpulent frame, harsh voice, and loud laugh," and vulgar traits such as "a liking for tobacco and spirits, and a relish for broad wit and low company." Her relatives and acquaintances whispered that as an infant she had been "put to nurse by a fat Irish woman." In a similar vein, physician Joseph Edcil Winters explained the "secretive disposition" of one youngster with reference to an Italian wet nurse. He . . . [also] reported that a medical student had told him that one of his brothers had been nursed by an Irish woman and exhibited "very decided Irish traits, which are so marked that they are noticed by all the friends of the family."[31]

A more scientific-sounding version of the same idea is that essences are located in one's DNA (a notion helped along, no doubt, by the folk-theory that racial essences are transmitted in seminal fluid). Although it has a veneer of scientific respectability, this DNA theory is only marginally less baseless than the theories about blood and milk, for, as we have seen, conventional racial categories are folk categories rather than scientific ones, and don't have any genetic justification.

Finally, there are more mystical explanations of the racial essence, framed in terms of the "spirit" or "soul" of a people—a nonphysical substance that somehow permeates them. I mentioned in Chapter Five that Hitler believed that races are "spiritual" rather than biological groups. However, irrespective of whether imagined as carried by blood, by genes, or by some spooky spiritual stuff, race is supposedly passed down from parents to their children. Recall the problem that confronted the Nazis, who were unable to find a surefire way to distinguish between Jews and Aryans. It was impossible to tell Jews from Germans just by looking at them, or by measuring their facial features, or by analyzing their blood chemistry. So, the guardians of racial purity decided to determine race by descent. A similar, albeit more extreme, criterion was used by whites in the southern United States to determine who was black. According to the "one-drop rule," a fair-skinned, blue-eyed blond could be considered "black" if they had even a single "drop" of "black blood" coursing through their veins, as determined by their ancestry.[32] In Brazil, the one-drop rule applies in reverse. A person can have dark skin, but if they have European ancestors, they may be considered "white." Racially mixed parentage creates a problem, because it's part of the logic of essences that they're all-or-nothing: they don't come in degrees. An object can't be only somewhat gold—it must be either gold, not gold, or gold combined with something else. By the same token, within a racially essentialist framework the offspring of mixed race parents are often conceived as a mixture, but not a compound, of the parental races (in this respect, the fact that we use the term "mixed race" rather than, say, "blended race," is quite revealing). This was the rationale for the one-drop rule that black trumps white (even today in the United States, a person with one dark-skinned parent is often described as "black"). Alternatively, the two essences may be imagined to vie with one another in a single person, producing a tragically divided or degenerate being.[33]

It appears that a race is *any group of people conceived as a natural human kind in virtue of sharing a heritable essence.* This explains why skin color, hair texture, and other phenotypic traits are neither necessary nor sufficient for individuating races and why ethnic groups like Italians and the Irish were once considered races (and in the contemporary United States, Hispanics are a de facto race). If I am right, the notion of race, *as it actually functions in human cognition and discourse,* is sometimes indistinguishable from notions of ethnicity, nationality, or even religious or political affiliation. Populations are often conceived as races even though they aren't labeled as such, because race isn't primarily about what people are *called*—it's about what they are *thought to be* (including, of course, what they think themselves to be). To avoid confusion, I will for the most part call these *ethnoraces*.[34]

Psychologists Maaris Raudsepp and Wolfgang Wagner provide an interesting example of ethnoracial thinking operating outside the bounds of conventional racial territory that also shows the connection between race and dehumanization. The small Baltic state of Estonia was part of the Soviet Union from 1918 until the latter disintegrated in 1991. Since achieving independence, Estonian nationalists have made efforts to denigrate ethnic Russians living there. The nationalists have articulated a vision of an Estonian essence that Russians—as inferior and alien beings—do not share. Raudsepp and Wagner have found that both native Estonians and Russian-Estonians "raise the issue of out-group characteristics that are supposedly incompatible with the in-group's essential attributes."

> Many of those who see themselves as native Estonians mention the Russian-ness . . . of the out-group, which is supposed to include Asian barbarism and communist mentality as expressed in this adapted version of the Estonian proverb "You may feed a wolf [*tibia*—a derogatory name for Russians] as long as you like; he will still look in the direction of the forest." Alternatively, members of the Russian group define Estonian-ness as being characterized by peasant barbarism and fascist mentality.

In effect, ideologues from both groups regard Estonians and Russians as distinct ethnoraces. As one would expect, it's lineage that determines whether someone has a Russian or an Estonian essence. Raudsepp and Wagner remark, "Only people descending from parents or grandparents who were Estonian citizens sixty years ago, were considered legitimate citizens of the new state . . . citizenship is defined by a proof of one's lineage." Consequently, there have been calls by Estonian nationalists for mass deportation of Russians, who are characterized as subhuman predators ("Cohabitation of human beings with wild beasts is not possible for long. And it is a crime not to send wild beasts back to their natural environment. . . ."), as well as discourse suggesting that Russian-ness is encoded in Russians' DNA ("The Mongol gene of robbing, killing, and hating of work has been coded into Russians.").[35]

This example illustrates something important about dehumanization. For dehumanization to occur the target group must first be essentialized. *They*, the others, must be seen as a distinct kind of person: not just superficially different, but radically so. This pattern is borne out by all of the cases of dehumanization that have been surveyed so far in this book. Think of the genocides described in Chapter Five—Namibia, Turkey, the Holocaust, Cambodia, and Darfur. Think of the slave trade, the conquest of the Americas, and the horrors of World War II. In *every* case, the perpetrators believed that the people whom they dehumanized were ethnoracially different from themselves.

Having defined the target population as an alien natural human kind, the second step on the road to mass violence is to attribute a subhuman essence to them. The enemy is no longer another kind of human being. Like wolves in sheep's clothing, our enemies are another species, lurking behind a mask of humanity. They are only apparently human for the same reason that Mr. Oreo was only apparently white, and they are really subhuman for the same reason that Mr. Oreo was really black.

FROM RACES TO SPECIES

The psychological transition that I've just sketched raises a further question. How, why, and under what circumstances does believing that a group has an ethnoracial essence get transformed into believing that it has a subhuman essence? To explain this shift, we need to look more closely into the psychological mechanisms that lie behind ethnoracial beliefs.

A lot of the research into the psychology of natural kinds comes from the study of folk-biological concepts—everyday intuitions about plants and animals. Anthropologists and psychologists have found that, in every culture studied, living things are classified in much the same way. According to Scott Atran, "In every human society people think about plants and animals in the same special ways."* Atran points out that "people in all cultures classify plants and animals into species-like groups that biologists generally recognize as populations of interbreeding individuals adapted to an ecological niche." These groups are called *generic species*. In all of the cultures that have been studied, generic species are seen as part of a broader taxonomic scheme—a nested hierarchy of categories that is strikingly similar to the Linnaean system used by modern biologists, and the generic species level is regarded as especially important. Atran remarks that:

> [T]here is a commonsense assumption that each generic species has an underlying causal nature, or essence, that is uniquely responsible for the typical appearance, behavior, and ecological preferences of the kind. People in diverse cultures consider this essence responsible for the organism's identity as a complex, self-preserving entity governed by dynamic internal processes that are lawful even when hidden. This hidden essence maintains the organism's integrity even as it causes the organism to grow, change form, and reproduce. For example, a tadpole and frog are in a crucial sense the same animal although they look and behave very differently and live in different places.

Finally, species classifications are used to draw conclusions about the natural world. Folk-taxonomies "provide an inductive framework

---

*This claim is less impressive than it sounds, as the folk-biological beliefs of relatively few cultures have been studied in depth.

for thinking about living things."[36] For example, if one porcupine is observed eating pine tree bark we suppose that all porcupines eat pine tree bark, and if we observe that one rattlesnake is poisonous we suppose that all rattlesnakes are poisonous. These commonsense assumptions are remarkably similar to the sorts of inductive inferences that biologists make all the time—probably because the impressive edifice of scientific biology was built on a deeply entrenched folk-biological foundation.

Even though biological essentialism is at the center of this way of thinking about the world, biological essentialism doesn't make scientific sense. There simply aren't any defining features—even at the genetic level—that all and only members of a species must share.[37] So, it can't be that our tendency to essentialize the natural world is based on observation. We can observe overt similarities between organisms, but we can't observe the essences assumed to account for these similarities. Folk-essentialism is a theoretical construction—an explanatory grid that we *impose* on the world rather than extract *from* it. Of course, it's not a formal theory—a theory of the sort that we derive from textbooks or other forms of didactic instruction. Rather, it's a folk-theory that arises in large measure from the cognitive architecture of the human mind.

Atran and others believe that folk-biological thinking is a domain-specific ability rather than a domain-general one. Domain-general abilities are patterns of thinking that apply across the board: we can bring them to bear on whatever subject matter we please. The capacity for abstract reasoning is a good example. A person who understands the basic principles of logic can use them to reason about whatever she chooses. They are as applicable to thinking about ships and snails and sealing wax as they are to cabbages and kings. Mathematical thinking is another. When we perform calculations, the kinds of things that we are calculating over don't matter. Two tigers plus three tigers equals five tigers just as surely as two volcanoes plus three volcanoes equals five volcanoes. In contrast, domain-specific cognition is limited to a specified area. It's fast, easily learned, and hard to shut off. Many cognitive neuroscientists believe that domain-specific thinking comes from the operation of special-purpose functional systems that were established in the human brain during the course of our evolution to deal with important, recurring problems faced by our prehistoric ancestors.[38]

Our remarkable ability to recognize and remember faces is a good candidate for a domain-specific ability. Faces are much easier to remember than names. How many times have you met someone and recognized that you'd met them before but couldn't remember their name? Research suggests that the human brain has a special knack for processing information about human faces. We are hypersocial animals, and the ability to keep track of members of our communities has always been very important for our well-being. Faces are the most reliable way to recognize individuals, and facial expressions are the best windows into their emotional states, so natural selection got to work and installed high-powered face-recognition software in our prehistoric ancestors' brains. This isn't just armchair speculation. There's plenty of evidence suggesting that the human brain contains a neural system specialized for face recognition. The brain handles face perception differently than it does ordinary object perception. When we look at faces, our brains attend to the total configuration of a face more than on its individual parts, which is why when faces are turned upside down they're much harder to recognize. On the developmental side, newborn babies prefer gazing at faces (or facelike objects) to looking at other things, and they learn to recognize individual faces from very early on. Most intriguingly, people suffering from a neurological disorder called *prosopagnosia* (or "face blindness") can recognize ordinary objects perfectly well, but can't recognize faces. This suggests that the brain system used to process information about faces is distinct from the one that processes other sorts of visual information.[39]

Likewise, there is evidence pointing to the existence of a folk-biology module—a domain-specific cognitive system specifically concerned with thinking about organisms. As we've seen, species essentialism is found across diverse cultures and emerges early in childhood. There is even a folk-biological equivalent to prosopagnosia. Damage to the left temporal lobe of the brain can knock out a person's capacity to recognize biological kinds, but has no effect on their ability to recognize artifacts, suggesting that there is a cognitive module for folk-biological thinking.[40] Studies of the psychological development of children also support the hypothesis that there are special-purpose cognitive systems specialized for folk-biological thinking. Yale University psychologist Frank Keil asked children a series of questions to explore their beliefs about natural kinds. For example, he asked them to imagine a scientist bleaching a tiger so that its stripes disappeared, and then surgically attaching a mane to its head so that it looked just like a lion. He then asked the kids whether the animal was a tiger or a lion. Kids who were younger than seven said that the animal was a lion, and justified their diagnosis by appealing to its appearance ("a tiger doesn't have hair on its neck"), but older children asserted that, in spite of its appearance, the animal was a tiger "because it was made out of a tiger." The same pattern of responses followed a story about a raccoon that was painted to resemble a skunk. The younger children considered it a skunk, and the older ones thought it was a raccoon that looked like a skunk.[41]

It would be hasty to conclude from these responses that young children don't essentialize species. Lions and tigers are quite similar, as are raccoons and skunks, so it may simply be that young children don't differentiate natural kinds in as fine-grained a manner as older kids do. This idea was supported by the results of another experiment. This time, the youngsters were told a story about a porcupine that was modified so as to be indistinguishable from a cactus, in what Keil calls a *cross-ontological category shift*. Even the youngest children were sure that it was still a porcupine despite its cactuslike appearance, as is illustrated by the following dialogue between experimenter and child.

> C: It's still a porcupine.
> E: What would happen when it wakes up? Will it be a cactus plant or will it be a porcupine?
> C: A porcupine.
> E: Why would you think that it would be a porcupine?
> C: It will look like a cactus, but it won't be one.
> E: Why not?

C: It'll be living more.
E: Are cactus living?
C: Yes, but it may be walking around.[42]

The young children may not reliably distinguish species as natural kinds, but they appear to attribute different essences to plants and animals.

Is this form of thinking specifically biological? Might it reflect their views about kinds in general? To investigate this question, Keil showed children pictures of a coffeepot and a birdfeeder with a similar shape, and read them the following story:

> There are things that look just like this (shows picture of coffeepot), and they are made in a big factory in Buffalo for people to make coffee in. They put the coffee grounds in here and then they add water and heat it all up on the stove and then they have coffee. A while ago some scientists looked at these things carefully and they found out they aren't like most coffeepots at all, because when they looked at them under a microscope, they found that they were not made out of the same stuff as most coffeepots. Instead they came from birdfeeders like this (points to picture of birdfeeder) which had been melted down and then made into these (points to picture of coffeepot) and when people were all done making coffee with these (points to picture of coffeepot), they melted them down again and made birdfeeders out of them. What are they, birdfeeders or coffeepots?

Both young and older children univocally claimed that the devices were coffeepots. When asked why, the responded like this:

C: A coffeepot.
E: Why?
C: Because it doesn't look like a birdfeeder. I wouldn't put birdseed in it. . . . I would put coffee in the coffeepot.

E: So even though this special object was made from birdfeeders that look just like this and when they were done with
them, they melt them down and make birdfeeders again,
you think the best name for it is what? A birdfeeder or a
coffeepot?

C: Coffeepot.

E: And your reason?

C: Because it doesn't look like a birdfeeder and I wouldn't put
birdseed in it.[43]

Essence trumps appearance when it comes to biological kinds, but
appearance calls the shots when distinguishing artificial kinds.

It should be clear by now that we tend to think about races along the
same lines as we think about species. Both races and species are presumed to be natural kinds defined by hidden essences passed down the
"bloodline" from parents to their offspring. Both are scientifically vacuous but intuitively compelling. Organisms may fail to manifest their
species-essence (even though stereotypical tigers have four legs and
stripes, three-legged tigers and a tiger without stripes still count as tigers) just as people can fail to manifest their racial essence (Janie and
Mr. Oreo were black, even though they looked white, and Yakov Bok
was Jewish, even though he didn't menstruate).

The striking similarity between patterns of folk-biological and racial
thinking suggests that racial thinking is domain-specific—i.e., that we
have an inbuilt tendency to divide the human race into discrete sub-
populations which we imagine as natural kinds. This tendency can be
resisted, or counteracted with education, but it is a pattern of thinking
into which people everywhere tend to slide, even when they know better.
Lawrence Hirschfeld, whom I introduced earlier in this chapter, is a
proponent of this hypothesis. His research provides a wealth of evidence
that even young children partition the human world into ethnoraces,
which they treat as natural kinds. Hirschfeld doesn't claim that small
children use the same racial categories as the adults around them do. It's
the *form* of ethnoracial thinking that spontaneously emerges during
childhood, rather than its specific content.

Hirschfeld and his coworkers designed and implemented several experiments to explore children's ethnoracial beliefs. In one, more than one hundred three-, four-, and seven-year-old children were presented with drawings depicting an adult and two children. The figures in the drawings represented either blacks or whites, as indicated by skin color, hair, and shape of nose and lips. Each person was wearing some sort of occupational uniform, and each had either a thin or a stocky physique. Each of the children was presented with three such drawings. In every drawing the children shared two characteristics with one another, but only one with the adult. For instance, one illustration portrayed a heavy-set black woman dressed in a nurse's uniform, with two children, a slender black girl dressed in a nurse's uniform and a stocky black girl dressed in ordinary clothes, while another showed a stocky black man in a policeman's uniform, a stout white child in a policeman's uniform, and a heavy-set black child in ordinary clothes.

Hirschfeld split the young experimental subjects into three groups, each of which had the task of matching the adult with one of the children. One group was asked whose parent the adult is, another group was asked which of the children represents the adult as a child, and the third group was asked which of the two children most closely resembles the adult. "The logic of the task is straightforward," he explains.

> In each triad, each of the comparison pictures shares two features with the target adult, but they share only one feature between themselves. One triad set contrasts race to bodily build, one contrasts race to occupation, and one contrasts occupation to bodily build. When asked to choose which of the comparison pictures is the target as a child, the target's child, or most similar to the target, children must decide which of the contrasted properties is most relevant. If children simply rely on outward appearances in making identity judgments, then they should be as likely to rely on one form of outward appearance as on another. Accordingly, they should choose at random. If they believe that one dimension

contributes more to identity than another, they should rely
on that dimension in making their choices.[44]

The results were impressive. The children's responses showed that
they believed that racial characteristics are more likely to be inherited
and to remain constant across a person's lifetime than either occupation or body build. Even the three-year-olds tended to think of races as
natural kinds, although their answers were the least consistent.

The experiment also shows that children don't group people into
races because they *observe* them to be especially similar. To kids, two
people wearing policemen's uniforms are just as similar as two people
with the same skin color. This suggests that children are inclined to
think *theoretically* about race—in much the same way that we tend
to think about species. Children implicitly believe that observable traits
like skin color are signs of a deep essence that unites the members of a
racial group under a single umbrella. Even three-year-olds think of racial attributes as "immutable, corporeal, differentiated, derived from
family background, and at least consistent with biological principles of
causality."[45]

Research also indicates that youngsters have the concept of race
before they have any understanding of the observable traits that racial categories are supposed to latch on to. This is most effectively
illustrated by looking at children's efforts to make sense of race in
their everyday lives.

> Ramsey . . . reports that a white three-year-old looked at a
> photograph of a black child and declared "His teeth are dif-
> ferent!" Then the subject "looked again, seemed puzzled
> and hesitantly said, 'No, his skin is different.'" At the age of
> four my daughter made a similar observation. She and I were
> stopped at a traffic light in France. She looked at an ethni-
> cally Asian family in the car next to us and then exclaimed
> that they looked like her friend Alexandre, a Eurasian child.
> I asked her in what ways she thought they looked like Alex-
> andre. She mulled over the question for a moment, staring

intently at the family as we waited for the light to change. Finally she said, "They all have the same color hair."

Hirschfeld goes on to remark:

> Of course she was right. Alexandre and the members of his family all had black hair—but so did I, at the time. More critically, the majority of the inhabitants of France have black hair whether they are ethnically French, Southeast Asian, or North African. The point to be taken from both stories is that preschoolers are aware of perceptual differences between members of racial groups and are aware that perceptual cues play a role in defining racial groups, but they do not appear to reflect on precisely which perceptual factors are important.[46]

Hirschfeld believes that there is a reciprocal relationship between our innate essentializing tendencies and the social dimension of racial categories. Because they're primed to essentialize, children quickly take adult categories on board. When children are told about racial categories (for example, that there are "white" and "black" people), they don't assume that members of these groups have a common appearance. For young children, factors like skin color or hair texture may indicate a person's race but they don't define it. They only gradually learn about the physical characteristics associated in their culture with races and come to integrate these social constructions in accord with their preexisting template.

The fact that ethnoracial thinking is already biologically tinged early in childhood is intriguing. But what, exactly, does it suggest? Why do our concepts of races and biological species have such an uncanny resemblance to one another? One possibility is that they're cut from the same cloth. Hirschfeld suggests that the human mind possesses a specialized cognitive module—which he calls the *human kinds module*—that uses the same general principles as the module responsible for folk-biological reasoning to draw inferences about human populations.

Just as the biological kinds module gives rise to folk-biological thinking, the hypothesized human kinds module produces a more-or-less universal form of folk-sociological thinking. The two mental systems operate independently of one another; both interpret their respective domains in terms of essences and natural kinds.[47]

ON THE ORIGIN OF PSEUDOSPECIES

Political enemies from western Kenya are called *nyamu cia ruguru*—animals from the west
—KOIGI WA WAMWERE, *NEGATIVE ETHNICITY: FROM BIAS TO GENOCIDE*[48]

The close psychological relationship between biological and social essentialism might suggest that we unconsciously use the cognitive machinery originally evolved to make sense of the biological world to make sense of the social world. That is, when confronted with human groups, we tend automatically to think of them along the same lines as we think of biological species. This idea is set out in an article entitled, "Are ethnic groups biological 'species' to the human brain?" by an anthropologist named Francisco Gil-White.

To get your mind around Gil-White's hypothesis, begin by considering the social world in which our prehistoric ancestors lived. Of course, it isn't possible to know for certain what the conditions of life were like, but there is more than enough evidence to allow us to make some well-informed conjectures. As far as we know, early humans lived in small, homogenous communities and rarely if ever encountered others who were physically unlike themselves. The overwhelming majority would never have met people with dramatically contrasting skin color, hair texture, or other phenotypic traits that we nowadays subsume under race. However, they almost certainly interacted with diverse cultural groups. It's very likely that these smallish bands were part of larger communities consisting of hundreds of individuals that anthropologists call "ethnies" or "tribes." Members of a tribe share a wealth of culturally transmitted beliefs, preferences, and rules of conduct—including the rule that one should mate only with members of the tribe ("normative endogamy"). Archeological evidence suggests that our Stone Age ancestors were organized into ethnies at least fifty thousand years ago, and possibly much earlier.[49]

All things being equal, it is much easier to deal with members of one's own tribe—people with a shared understanding of a common way of life, who speak the same language and adhere to the same norms and values—than it is to engage in social exchange with outsiders. Social interaction across tribal boundaries is a minefield, rife with opportunities for misunderstanding, conflict, and—at the extreme—danger. Given this, it was advantageous for tribal groups to adopt conspicuous symbolic paraphernalia such as dress and body paint, scarification and jewelry, as well as forms of speech and ritualistic behaviors such as greeting rituals, eating rituals, dances, and religious practices to demarcate themselves from all others. Ethnic markers worked like cultural traffic lights to regulate the flow of social exchange and keep it as much as possible within the group. The mortal significance of ethnic markers is beautifully illustrated by a story in the biblical book of Judges, which highlights how something as seemingly trivial as the pronunciation of a word can be used to demarcate friend from foe. The story, which is set in the eleventh century BC, concerns a war between the tribes of Ephraim and Gilead. Defeated, the surviving Ephraimites try to retreat across the river Jordan to return to their homeland, but the men of Gilead have anticipated this move and turned the fords into checkpoints. Trapped, the Ephraimites try to deceive the sentries by disguising themselves as men of Gilead, but their attempt is foiled by a simple test.

> Whenever a survivor of Ephraim said, "Let me cross over," the men of Gilead asked him, "Are you an Ephraimite?" If he replied, "No," they said, "All right, say 'Shibboleth.'" He said, "Sibboleth," because he could not pronounce the word correctly, they seized him and killed him at the fords of the Jordan. Forty-two thousand Ephraimites were killed at that time.[50]

Gil-White speculates that the growing insularity of ethnic communities, and the proliferation of ethnic markers setting them off from one another, created the illusion that ethnic identity is biologically inherited. You may recall that Erik Erikson captured this idea in his notion of cultural pseudospeciation.

> At its most benign . . . "pseudo" means only that something happens to appear to be what it is not; and, indeed, in the name of pseudospecies man could endow himself with pelts, feathers, and paints, and eventually costumes and uniforms— and his universe with tools and weapons, roles and rules, with legends, myths, and rituals, which served to bind his group together and endow its unique identity with that super-individual significance which inspires loyalty, heroism, and poetry.[51]

Erikson's remarks on pseudospeciation were very sketchy. Gil-White fills the story out with a plausible account detailing how pseudospeciation may have occurred. His idea is that once ethnic groups became consolidated, prehistoric humans began to respond to members of alien groups as though they were separate species. This happened because ethnic communities started to "look" like biological species to the human brain. Think about it. Ethnic communities adopted forms of display such as clothing and body paint that made them appear very different from one another. They also adopted different forms of behavior— especially speech and cultural rituals. Finally, and perhaps most significantly, they restricted marriage to other members of the tribe, which led to the consequence that ethnic membership was determined by descent. Members of each tribe possessed what amounted to a "cultural phenotype" that was handed down from parents to their offspring which echoed the reproductively transmitted biological phenotypes found in nature. The stage was now set for ethnic groups to trigger the cognitive module for intuitive biology, and thus cause the brain to process information about ethnic groups as though they were distinct biological species.

"Over time . . . ," Gil-White suggests, "the brain evolved to improve the fit and make ethnies part of the 'proper domain' of the living kind module, completing the *exaptation*."[52] *Exaptation* is a biological term for features of organisms that take on functions that evolution didn't originally adapt them for. Penguins' wings are a nice example. Ancestors of modern-day penguins used their wings for flying, but their descendents embraced a semi-aquatic lifestyle, causing the ancestral wings to change their function.* The wings were modified to become flippers. There isn't a sharp line to be drawn between adaptation and exaptation—in fact, exaptation might just as well be called "re-adaptation."

Gil-White's hypothesis, then, is that once distinctive human cultures emerged, the biological kinds module took on an entirely new function. Its domain of operation expanded to include human kinds, and intuitive folk-biology gave birth to intuitive folk-sociology. As a result, ethnic groups look like biological species to the human brain. Gil-White calls this the Ugly Duckling hypothesis.

> The Ugly Duckling hypothesis predicts that categories that look like a species (i.e., meet the brain's "input criteria" of a species) will tend to be essentialized, especially when the perceptions of descent-based membership and category-based endogamy, in particular, are strong. A corollary is that inductive generalizations of nonobvious properties—which essentialist intuitions motivate—will be more easily made in categories that look like species.[53]

Hirschfeld and Gil-White propose rival explanations of why folk-sociology looks so much like folk-biology, but neither of them uses their theory to address dehumanization—the place where folk-sociology and folk-biology intersect—and it's useful to consider how each theory could be used to explain it. Hirschfeld's hypothesis suggests that we move so

---

*The full story is a lot longer. Birds' wings were originally exapted from forelimbs, and their feathers were apparently exapted from reptile scales.

easily from racism to dehumanization because of the isomorphy between folk-sociological and folk-biological thinking. The psychological similarity between racial and biological thinking might explain why we often mix up the former with the latter, just like we sometimes confuse similar-sounding words. Now, add a dash of motivation to the mix. Because thoughts about ethnoracial groups have a deep resonance with thoughts about biological species, people's minds naturally turn to thoughts about the latter when they want to denigrate the former. Because derogatory thoughts are the driving force, hated or despised species are unconsciously selected to represent them. However, Hirschfeld's theory doesn't explain *why* folk-biology and folk-sociology conform to the same pattern. Gil-White's account fills in the gap, and in doing so suggests that the tie between folk-biology and folk-sociology is more intimate than structural isomorphism would allow. If folk-sociology is built on folk-biological foundations, then folk-biology is the default position. This might explain why ethnoracial categories so readily collapse into biological ones, but not vice versa.

SUMMARY

As the trajectory in this chapter has been fairly complex, and perhaps more than a little confusing, I think that it will be useful to conclude by summarizing the main points.

1. We intuitively carve humanity up into natural human kinds or "ethnoraces" modeled on biological species. We have a "folk-sociological" theory that strongly resembles our "folk-biological" theory. These ways of thinking about the world are natural and compelling, and widely distributed across cultures, even though they are inconsistent with the scientific picture.

2. The form of ethnoracial thinking is innate, while its content is determined by cultural beliefs and ideologies.

3. Ethnoraces are believed to share an essence that defines the

kind—there is a mysterious "something" that makes one African American, or Jewish, or Tutsi, or Irish. This essence is imagined as being somehow "inside" a person, but distributed throughout them rather than localized in a particular part. It is thought to be carried by bodily fluids—especially blood.

4. Knowing a person's ethnorace is supposed to allow one to make inferences about his or her nonobvious properties.

5. A person's ethnorace is considered to be a necessary, and therefore unalterable, characteristic. It remains constant at every possible world where they exist.

6. Ethnoracial essence is taken to be responsible for stereotypical characteristics of natural human kinds. These attributes may or may not be expressed. If they are unexpressed, an individual who is a member of the kind may appear not to be a member of the kind.

7. Folk-sociological thinking may be the product of a domain-specific cognitive module: either a distinct "human kinds" module (Hirschfeld) or an extension of a "living kinds" module (Gil-White).

8. Ethnoraces are the objects of dehumanization. First, a population is imagined as a natural human kind with a common essence, and second, their common essence is imagined to be a subhuman essence.

Having gotten this far, we are now positioned to put in place the last few pieces of the puzzle of dehumanization. Neither Hirschfeld nor Gil-White set out to analyze dehumanization, but their observations about the similarity between folk-biological and folk-sociological thinking is crucial for accomplishing this.

7

THE CRUEL ANIMAL

Ever since Darwin, the dominant tendency in comparative cognitive psychology has been to emphasize the con-
tinuity between human and nonhuman minds as "one of
degree, and not of kind."

—DEREK PENN, KEITH HOLYOAK,
AND DANIEL POVINELLI, "DARWIN'S MISTAKE"[1]

I'll teach you differences.

—WILLIAM SHAKESPEARE, *KING LEAR*[2]

FOR CENTURIES, PHILOSOPHERS AND THEOLOGIANS have pon-
dered what it is that sets us apart from the rest of the animal kingdom.
Aristotle thought it was rationality. To others, it was the possession of an
immortal soul or having been fashioned in the image of God. Mark
Twain, who loved to deflate human arrogance, had his own suggestion.
In an 1896 essay entitled "Man's Place in the Animal World," he argued
that Man belongs at the bottom of the great chain of being, rather than
near the top. Twain, an ardent Darwinian, suggested acerbically that
his observations of human nature obliged him "to renounce my alle-
giance to the Darwinian theory of the Ascent of Man from the Lower
Animals" in favor of "a new and truer one . . . the Descent of Man from
the Higher Animals." Our descent is simultaneously biological and
moral. We are, Twain believed, the only species capable of immorality.

Man is the Cruel Animal. He is alone in that distinction. . . .
Man is the only animal that deals in that atrocity of atrocities, war. He is the only one that gathers his brethren about
him and goes forth in cold blood and calm pulse to exterminate his kind. He is the only animal that for sordid wages
will march out . . . and help to slaughter strangers of his
own species who have done him no harm and with whom
he has no quarrel. . . . And in the intervals between campaigns he washes the blood off his hands, and works for
"the universal brotherhood of man" with his mouth.[3]

Although Twain's essay wasn't intended as a scientific or philosophical treatise, it makes a number of points that merit serious attention.
The passage that I've just quoted makes two important claims about
human uniqueness. The first one is that humans are the only animals
that are cruel. The point isn't that *all* humans are cruel (i.e., that a person who isn't cruel isn't human). Rather it's that humans are the only
animals that *have it in them* to be cruel. In other words, Twain hinted at
the idea that there is something about the way that human nature is
configured that makes it possible for us to be cruel. Twain's second
claim, which is closely bound up the first, is that humans are the only
animals that wage war on one another.

Notice that Twain didn't propose humans are the only animals that
kill members of their own species, or kill them en masse, but he implies mass killing is war only if certain conditions are fulfilled: war is
calculated slaughter ("cold blood and calm pulse") that's not motivated by a personal desire for retribution ("slaughter strangers . . . who
have done him no harm and with whom he has no quarrel") and is
undertaken for personal gain at the behest of a third party ("for sordid
wages"). This comes pretty close to how many present-day anthropologists define war.

In this chapter, I'm going to defend the proposition that *Homo sapiens* are the only animals capable of cruelty and war. I'm going to explore where this leads, and use it to develop a more detailed explanation
of why dehumanization causes moral disengagement.

IS WAR UNIQUELY HUMAN?

There are plenty of intelligent, scientifically literate people who reject the idea that humans are the only animals that go to war. Their response to Twain's essay might be to claim that, although Mark Twain was a brilliant man and a great writer, he was inevitably a prisoner of his time, and couldn't have anticipated the major discoveries about animal behavior that would be made over the next century or so—discoveries that prove him wrong. What discoveries? Our interlocutor would cite two, involving ants and chimpanzees.

Let's begin with ants.

Ant colonies sometimes attack other colonies of the same species. These attacks are sometimes described as wars. That's why Edward O. Wilson and Bert Hölldobler have a chapter entitled "War and Foreign Policy" in their book *Journey to the Ants*. Here's how the chapter begins.

> The spectacle of the weaver ants, their colonies locked in chronic border skirmishes like so many Italian city-states, exemplifies a condition found throughout the social insects. Ants in particular are the most aggressive and warlike of all animals. They far exceed human beings in organized nastiness; our species is by comparison gentle and sweet-tempered. The foreign policy aim of ants can be summed up as follows: restless aggression, territorial conquest, and genocidal annihilation of neighboring colonies whenever possible. If ants had nuclear weapons, they would probably end the world in a week.[4]

Wilson and Hölldobler are two of the world's leading authorities on ants, so we can trust that they've got their facts straight. But it's important to bear in mind that they are using a colorful idiom for the purpose of engaging a popular readership. They know better than anyone that ant "war" is instinctual behavior mainly controlled by hard-wired responses to chemical signals, that ants don't really have foreign policy, and that their "genocidal annihilation of neighboring colonies" bears little relation to Auschwitz or Rwanda. Obviously, if ants had nuclear weapons they'd crawl on them rather than use them to end the world in a week! The "wars" waged by ants are *metaphorical* wars, not real ones.

A stronger and more compelling case can be made for chimpanzee war. Chimpanzees live in communities called "troops" that have what is called a fission-fusion organization. Each day, subgroups and lone individuals wander off into the forest to forage for fruit or hunt for small animals (fission), and at the end of the day they come back together and reconstitute the group (fusion). They are fiercely xenophobic animals with large, overlapping ranges, so there's always a chance that a foraging party will encounter chimps from a neighboring community. These meetings spark hostility. When two groups meet, nothing more serious occurs than noisy threat displays, but if a party of males happens upon a solitary stranger, they are likely to kill him.

These chance encounters aren't the only form of intergroup violence. Sometimes parties of six or so males make deep incursions into their neighbors' territory, apparently searching for individuals to kill. If they spot a lone male, a lone older female, or a male-female pair, they will ferociously attack and kill it. Young females are usually spared, and are incorporated into the attackers' troop.

These attacks are known as *raids*. When Jane Goodall and her team first began to study the chimpanzees of Tanzania's Gombe National Park, they were all members of a single troop, which the scientists called *Kasekela*. In 1971, they noticed that a subgroup of seven adult males, three adult females, and their offspring had begun to split off to form a second troop. A year later they were completely separate from the Kasekela community and had established their own independent territory. The scientists called them the Kahama group. Kahama now occupied part of Kasekela's old range, leading to tensions between the two communities. Over the next couple of years, aggressive confrontations became more and more frequent. Then, in 1974, a party of six Kasekela males traveled south to Kahama territory and killed an adult male, and over the next four years, all of the Kahama adults, and five

Kasekela females, were either killed or had disappeared. The juvenile Kahama females were absorbed into the Kasekela community.

Chimpanzees aren't vegetarians. They supplement a diet of fruit with the raw flesh of other mammals. They're especially fond of red colobus monkeys. They hunt monkeys in groups, and almost all the hunting is done by males. Richard Wrangham and Dale Peterson explain that monkey hunts are occasions of frenzied excitement.

> The forest comes alive with the barks and hoots and cries of the apes, and aroused chimpanzees race in from several directions. The monkey may be eaten alive, shrieking as it is torn apart. Dominant males try to seize the prey, leading to fights and charges and screams of rage. For one or two hours or more, the thrilled apes tear apart and devour the monkey. This is blood lust in its rawest form.[5]

When chimpanzees embark on a raid, their behavior resembles a monkey hunt. They're out for blood—but this time it's the blood of a member of their own species.

> Based on chimpanzees' alert, enthusiastic behavior, these raids are exciting events for them. . . . During these raids on other communities the attackers do as they do while hunting monkeys, except that the target "prey" is a member of their own species.

When a chimpanzee captures a colobus monkey, the monkey is immediately killed—usually by flailing it against a tree—and eaten. Raiding chimpanzees don't eat their quarry, but they attack it with utmost ferocity. It's not just killing—it's overkill. Wrangham and Peterson report that "their assaults . . . are marked by a gratuitous cruelty—tearing off pieces of skin, for example, twisting limbs until they break, or drinking a victim's blood—reminiscent of acts that among humans are unspeakable crimes during peacetime and atrocities during war."[6]

These attacks (as well as observations of female chimpanzees devouring infants) were first observed by Jane Goodall's team in Gombe National Park. Goodall was horrified by what she saw, which, she reports, "changed forever my view of chimpanzee nature . . . ."

> For several years I struggled to come to terms with this new knowledge. Often when I woke in the night, horrific pictures sprung unbidden to my mind—Satan, cupping his hand below Sniff's chin to drink the blood that welled from a great wound on his face; old Rodolph, usually so benign, standing upright to hurl a four-pound rock at Godi's prostrate body; Jomeo, tearing a strip of skin from Dé's thigh; Figan, charging and hitting, again and again, the stricken, quivering body of Goliath, one of his childhood heroes; and, perhaps worst of all, Passion gorging on the flesh of Gilka's baby, her mouth smeared with blood, like some grotesque vampire from the legends of childhood.[7]

The "war" between the Gombe chimps was the first time that this sort of behavior was observed. But it wasn't the last. Subsequent observations of interactions between troops of chimpanzees in at least seven other African regions have demonstrated that lethal violence isn't an isolated or freakish occurrence, but a normal albeit infrequent aspect of chimpanzee life.[8]

These facts about chimpanzees invite a comparison with human societies. Many human societies also conduct raids against their neighbors—in fact, it's the most common form of intercommunity violence between tribal groups. The Yanomamö of northern Brazil and Venezuela are especially well known for their raiding, largely through the work of social anthropologist Napoleon Chagnon, who studied the Yanomamö for more than thirty years. According to Chagnon, when Yanomamö warriors launch a raid, they first develop a plan of attack. Next they split into small groups of four to six men, so that each group can cover the other when making a retreat. The raiders approach the village before daybreak and hide in the bush near paths that lead to sources of drinking water. Their goal is to ambush a single person, and then leave the area before the body is discovered. They also abduct young women whenever possible. The fate of captured women isn't pretty. First, they are raped by all the men in the raiding party, and then, when the raiding party gets back to their village, other men rape her as well. At the end of this ordeal, she's awarded to one of the men as a wife.

Yanomamö raiding has some striking similarities to the raiding behavior of chimpanzees. In both cases a group of males stealthily enter enemy territory to ambush an individual or small group and, if possible, abduct fertile females. Once the job is done, they quickly return to their base. This similarity may reflect the biological relationship between humans and chimpanzees. Chimpanzees are our closest living relatives. Around six and a half million years ago both species had an ancestor in common, and since that time the two lineages have gone their separate ways. Chimpanzees seem to have evolved more slowly than humans, so they're more similar to the common ancestor than we are. Because of this, we can use chimpanzees as a model for some of our earliest ancestors. Of course, chimps aren't *exactly* like our earliest ancestors, but they're probably close enough to allow us to draw some tentative conclusions about our past. Although we can't be certain, it's reasonable to suppose that the social behavior of chimpanzees mirrors the social behavior of our common ancestor, which suggests that the ancestral ape lived in communities with a fission-fusion structure, was territorial, and was hostile toward neighbors of the same species.

Chimpanzee raids sometimes promise immediate rewards, such as food or mating opportunities. But that's not always the case. Chimpanzees often attack their neighbors when food is abundant in their own territory, and there are no females around for them to abduct, and it would be absurd to think that killing is part of a long-term strategic plan. So, this sort of killing must be instinctive. It must be that male chimpanzees simply have the urge to eradicate outsiders when they can do so without putting themselves at risk (chimps will only attack if they outnumber the opponent by a factor of at least three to one). Wrangham calls this the *dominance drive* hypothesis. The idea is that evolution installed this drive in male chimpanzees because killing outsiders puts the neighboring group at a disadvantage in competition for resources. This is precisely what happened in the conflict between the Kasekela and Kahama communities; the Kasekela group incrementally weakened the Kahama group by picking them off over a period of four years. In the end, Kahama was destroyed, their territory was annexed, and the Kasekela males got access to young Kahama females.

Wrangham goes on to suggest that the same approach may explain the evolution of lethal violence between human communities, culminating in war. There are two steps to his argument. The first is to make a connection between the xenophobic tendencies and intergroup hostility that are so evident among human beings, and the rampant xenophobia and intergroup hostility of nonhuman primates. Given that we *Homo sapiens* are primates, the obvious conclusion to draw is that our unlovely tendencies are biologically rooted, and that we inherited them from our common ancestor with the chimpanzees. The next step is to argue that evolution fashioned primates—including human ones—to behave in this way because picking off members of neighboring communities gives one's own community a competitive advantage. This, in turn, has important implications for understanding the male psyche. Men have, by nature, a "demonic" side. As Wrangham remarks, "there has been selection for a male psyche that, in certain circumstances, seeks opportunities to carry out low-cost attacks on unsuspecting neighbors."

> The psychological mechanisms that would make such a complex function possible have not been studied, but a partial list might include: the experience of a victory thrill, an enjoyment of the chase, a tendency for easy dehumanization [or "dechimpization" . . . i.e., treating nongroup members as equivalent to prey], and deindividuation (subordination of own goals to the group, ready coalition formation, and sophisticated assessment of power differentials).[9]

I've juxtaposed chimpanzee raiding with Yanomamö raiding to bring out their similarities. But we shouldn't let these blind us to their differences. So, let's consider the differences.

One significant difference has to do with forethought. Chimpanzees' behavior suggests that they plan, but only in the sense of adjusting their attack behavior to suit the circumstances. This is moment-to-moment planning involving little forethought. In contrast, the Yanomamö make elaborate preparations long before embarking on a raid. This goes hand in hand with another sort of difference. Yanomamö raiding is embedded in an intricate network of traditional norms, beliefs, and customs. As Rutgers University anthropologist R. Brian Ferguson remarks:

> Yanomami conceptualize, and so act out, war through complex cultural constructions involving classifications of social and moral distance, ideas about physical and supernatural aggression, and a pattern of symbolic and ritual exchanges. These are "essential constitutive dimensions of warfare as a social institution" and any analysis that leaves them out is incomplete, impoverished.[10]

The rich cultural dimension of Yanomamö raiding has no echo in chimpanzee society. The cultural dimension is especially evident in the ways that the Yanomamö prepare for their raids. Chimpanzee raids are instigated on the spur of the moment. Wrangham and Peterson tell us that:

> A raid could begin deep in the home area, with several small parties and individuals of the community calling to each other. Sometimes the most dominant male—the alpha male—charged between the small parties dragging branches, clearly excited. Others would watch and soon catch his mood. After a few minutes they would join him. The alpha male would only have to check back over his shoulder a few times.[11]

In contrast, Yanomamö preparations are quite elaborate. In one example described by Chagnon, the festivities began the day before with a ceremonial feast. Then, after the feast, a grass dummy representing the enemy was set up in the village center. The men ritually attacked it, killing their enemy in effigy. The man painted themselves black and crept stealthily around their own village, pretending to search for the enemy's tracks, and then after firing a volley of arrows, screamed and ran out of the village in a feigned retreat. After this, the warriors retired to their hammocks, and resumed the ceremony after darkness fell. One by one, they marched to the center of the village, clacking their arrows against their bows and making animalistic sounds. This lasted about twenty minutes. Once it was finished, the fifty-man raiding party assembled at the village center.

> When the last one was in line, the murmurs among the women and children died down and all was quiet in the village once again. . . . Then the silence was broken when a single man began singing in a deep baritone voice: "I am meat hungry! I am meat hungry! Like the carrion-eating buzzard I hunger for flesh!" When he completed the last line, the rest of the raiders repeated his song, ending in an ear-piercing, high-pitched scream. . . . A second chorus led by the same man followed the scream.

After several more choruses, during which the warriors worked themselves into a rage, they moved into a tight formation with their weapons held aloft.

> They shouted three times, beginning modestly and then increasing their volume until they reached a climax at the end of the third shout: "Whaaa! Whaaa! WHAAA!" They listened as the jungle echoed back their last shout, identified by them as the spirit of the enemy. They noted the direction from which the echo came. On hearing it, they pranced about frantically, hissed and groaned, waving their weapons . . . and the shouting was repeated three more times. At the end of the third shout of the third repetition,

the formation broke, and the men ran back to their respective houses, each making a noise—"Bubububububububu!"—as he ran. When they reached their hammocks, they all simulated vomiting, passing out of their mouths the rotten flesh of the enemy they had symbolically devoured in the line-up.

At dawn, the women collected food, and arranged it in parcels for the outward-bound raiders to collect. And then the men emerged, painted black, for a final ritual assembly at the village center, before marching out of the village.[12]

There's another noteworthy difference that reveals something important about an important difference between humans and chimpanzees. Cast your mind back to the "war" between the two groups of Gombe chimpanzees. Richard Wrangham, who watched the conflict take its deadly course, remarks that, "Horrifying though these events were, the most difficult aspect to accept was not the physical unpleasantness but the fact that the attackers knew their victims so well. They had been close companions before the community split."

> It was hard for the researchers to reconcile these episodes with the opposite but equally accurate observations of adult males sharing friendship and fun: lolling against each other on sleepy afternoons, laughing together in childish play, romping around a tree trunk while batting at each other's feet, offering a handful of prized meat, making up after a squabble, grooming for long hours, staying with a sick friend. The new contrary episodes of violence bespoke huge emotions normally hidden, social attitudes that could switch with extraordinary and repulsive ease. We all found ourselves surprised, fascinated, and angry as the number of cases mounted. How could they kill their former friends like that?[13]

The fact that these apes turned against their old companions suggests that chimpanzees understand the concepts *us* and *them*, and that in the chimpanzee mind the division between the two is rigid, static, and biologically driven. As Goodall remarks:

> Chimpanzees . . . show differential behavior toward group and non-group members. Their sense of group identity is strong and they clearly know who "belongs" and who does not: non-community members may be attacked so fiercely that they die from their wounds. And this is not simple "fear of strangers"—Members of the Kahama community were familiar with the Kasekela aggressors yet they were attacked brutally. By separating themselves, it was as though they forfeited their "right" to be treated as group members.[14]

To delve a little deeper into this, let's set aside chimpanzees for a moment and think once again about ants. Ants are highly social, and extremely xenophobic. When an ant wanders into the wrong colony, the newcomer isn't welcomed with open antennae. At best she's tolerated and at worst she's immediately killed. She may also be shunned, threatened, or physically harassed. What's going on here? Do ants have the concepts *us* and *them*?

No, they don't.

It turns out that ants' behavior is controlled by hard-wired responses to chemical signals. Every ant colony has a distinctive scent, and when two ants meet, they sample each other's scent. If a stranger to the colony has a strongly alien smell, this triggers attack behavior, but if she smells only slightly odd, she's allowed to survive. If she's allowed to hang around for a couple of weeks, and picks up enough of the colony's scent, she's then treated as an equal. Ant xenophobia is all about chemistry; concepts don't come into it at all.[15]

For ants, being one of *us* is nothing above and beyond having a certain chemical property. With chimpanzees, things are completely different. When the Kahama chimps split off from Kasekela, these animals didn't change in any way. They still looked the same and smelled the same. Their personalities remained the same, and their old Kasekela friends could still recognize them as individuals. But *something* changed.

By changing their location, the Kahama chimpanzees crossed over an invisible boundary. This wasn't just a geographical boundary, it was a conceptual boundary as well—a boundary in the minds of the Kasekela chimps.

Chimpanzees seem to distinguish between the troop and the individuals that compose it, but they have an inflexible notion of the difference between friend and foe: members of the troop, the local breeding group, fall into the former category, and all other chimpanzees fall into the latter. In the chimpanzee mind, *us* equals troop members, so when an individual leaves the troop it thereby leaves the category of *us* and becomes an object of hostility. One of the more striking characteristics about humans, which chimpanzees completely lack, is our capacity to form alliances between groups. This is made even more arresting by the fact that allies are often former enemies. Human beings have a unique ability to unite disparate groups under the conceptual umbrella of a more inclusive *us*. Shifting patterns of alliance and estrangement, of inclusion and exclusion, characterize human societies everywhere. We can make peace, whether transient or long-lasting, with our enemies. Chimpanzees can't. Unlike chimpanzees—or any other nonhuman animal for that matter—humans live in what Cornell University sociologist Benedict Anderson has aptly named *imagined communities*— communities constituted and bounded by our concepts of them.[16]

Toting up the balance sheet, I think it's clear that the differences between Yanomamö and chimpanzee raiding vastly outweigh their similarities. Given the gulf that exists between human and chimpanzee patterns of violence, it seems misleading to refer to the latter as war. But it's also implausible to claim that there's no connection between the two—and perverse to assert that the human propensity of intergroup violence has nothing to do with the xenophobic behavior of our primate cousins.

I think that the most balanced assessment is that our primate heritage has left human beings (especially human males) with a disposition for violent aggression against outsiders, and that this is a necessary but not a sufficient condition for raiding, war, and other cruel and lethal cultural practices.

CRUELTY

This was a very innocent planet, before those great big
brains.

—KURT VONNEGUT, *GALÁPAGOS*[17]

There can be no doubt that the difference between the
mind of the lowest man and the highest animal is immense.

—CHARLES DARWIN, *THE DESCENT OF MAN*[18]

It's tempting to think about nonhuman animals as though their minds
are just simpler versions of our own. Beatrix the cat is busy stalking a
mouse that's nibbling bread crumbs on the dining-room floor. As she
moves slowly forward, freezing like an elegant feline statue after every
few steps, it's tempting to suppose that she *believes* that there's a mouse
nibbling bread crumbs on the dining-room floor—tempting, but wrong.
To have this belief, Beatrix would need to grasp the concepts *mouse,
dining room, bread crumb,* and *floor* (you can't conceive of a mouse on
the floor unless you understand what a mouse is and what a floor is).
Beatrix may have a belief about the animal on the dining-room floor
as she creeps up on it (I think she probably does), but they can't be anything like the human ones that we're inclined to attribute to her.

Anthropomorphizing the animal mind can lead to a lot of confusion
when talking about animal behavior and it doesn't help that many of
the terms that scientists use to describe animal behavior have different
meanings in everyday discourse. Take the concept of a *threat display.*
In everyday language, the word *threat* describes deliberate actions
undertaken to intimidate another person. You can threaten someone by
pointing a gun at them, shouting insults, or promising to do them harm.
But in each of the cases, your intentions matter. Suppose you were visiting Romania and tried to order a cup of coffee in garbled Romanian,
but ended up saying "I'm going to bite you" to the terrified waiter. This
wouldn't be a threat (even though the waiter would think it was)

because it didn't express an intention to intimidate. The concept of a threat display, as used by ethologists, has nothing to do with animals' intentions. Threat displays are just stereotyped patterns of behavior that have the function of scaring other animals away or making them back down. But if you're not careful, it's easy to slip into thinking anthropomorphically about threat displays. For instance, you might assume that Beatrix makes her fur stand on end when she meets the neighbor's cat because she believes that doing this will scare the neighbor's cat away.

Psychological research suggests that the human brain is wired for anthropomorphic thinking, what Hume called "a universal tendency among mankind to conceive all things like themselves."[19] University of Arizona anthropologist Stewart Guthrie confirms that we are incorrigible anthropomorphizers. "Faces and other human forms seem to pop out at us from all sides," he writes. "Chance images in the clouds, in landforms, and in inkblots present eyes, profiles, or whole figures. Voices murmur or whisper in wind or waves. We see the world not only as alive but also as humanlike." The tendency to perceive human intentions everywhere is even more pervasive than illusions of the human form.

> Nothing is so important to us as other humans. Because we are preoccupied with each other, we are sensitive to any possible human presence and have tolerant standards for detecting it. Mostly unconsciously, we fit the world first with diverse humanlike templates. Our preoccupation with a human prototype guides perception in daily life. We attend to what fits the humanlike templates and temporarily ignore what does not.[20]

Because chimpanzees are so obviously similar to us, it is extremely easy to attribute humanlike mental states to them. Earlier in this chapter, I quoted a remark by Wrangham and Peterson about "gratuitous cruelty" of raiding chimpanzees. I may be wrong, but I doubt very much that Wrangham and Peterson meant to say that chimpanzees are cruel *in the same sense* that human beings are cruel. I think that they used this phrase to engagingly express the idea, in a book written for a general audience, that when chimpanzees attack one another, they inflict more pain and damage than is necessary. But the phrase is ambiguous, and can be read as saying that chimpanzees are cruel in the same sense that humans are cruel.

This is something that Jane Goodall wondered about. Goodall confesses that prior to the discovery of the darker side of chimpanzee behavior, she "believed that chimpanzees . . . were rather 'nicer' than us." She was deeply disturbed by the new revelations and it took a while for her to conclude that "although the basic aggressive patterns of the chimpanzees are remarkably similar to some of our own, their comprehension of the suffering they inflict on their victims is very different from ours. . . . But only humans, I believe, are capable of deliberate cruelty—acting with the intention of causing pain and suffering."[21]

Although Goodall doesn't explain how she reached this conclusion, I think that there are very good reasons to accept it, and that these reveal something important about dehumanization. To get at these reasons, we need to give some thought to the nature of cruelty.

Like many other writers on the subject, Goodall understands cruelty as the deliberate infliction of pain and suffering. It's true that cruelty often involves deliberately inflicting pain and suffering, but it doesn't always. Shooting a person in the head at point blank range is cruel, even though it kills them instantaneously and therefore causes them no pain. Deliberately breaking the fingers of a person in a permanent vegetative state is cruel. It can even be cruel to give someone what they want: persistently offering cookies to a morbidly obese person is cruel, even though it gives them pleasure. On the other side of the coin, it's also possible to deliberately inflict pain and suffering on someone without thereby treating them cruelly. A physician who gives an injection to a child causes the child pain, but isn't being cruel.

Rewording the definition to say that cruel acts are undertaken for the purpose of causing pain doesn't help very much because it doesn't cover cases of shooting a man in the head and breaking the fingers of the comatose patient. We're going to have to look elsewhere to uncover the meaning of cruelty.

It seems more promising to think of cruelty as deliberately causing *harm*.* You can harm someone without causing them pain, and you can cause someone pain without harming them. Let's adopt this as a working definition of cruelty, and see where it leads. Obviously, to deliberately harm someone, you've got to be conscious of what you're doing; you've got to be aware that you want to harm someone, and also be aware that what you're doing or planning is harmful to them. In short, you've got to be able to reflect on your own actions and psychological states. You've also got to have the relevant concepts—for instance, the concept of *harm*, and you've also got to be able to tell the difference between things that can be harmed and things that can't (more about that in a moment). Now, chimps are very smart, but they're not *that* smart. There's no reason to suppose that they're able to reflect on their own intentions or that they can grasp sophisticated concepts like *harm*. So it looks like Jane Goodall was right. Chimpanzees can't be cruel.[22]

This isn't the end of the story about harm. I want to take it further, to explore the nature of human cruelty. In doing so, we'll not only come to appreciate part of what makes human nature unique, we'll get a deeper understanding of what it is about us that makes us capable of dehumanization.

Human beings use an array of concepts to make sense of the world around us. Without them, the world would be, in William James's famous words, a "buzzing, blooming confusion."[23] Our concepts come in several varieties. Some of them are descriptive—that is, they purport to represent objective features of the world. The concepts *red*, *big*, *hairy*, and *liquid* are all descriptive. Others are evaluative. Evaluative concepts pertain to the value that we give to things, and include notions like *good*, *bad*, *beautiful*, and *disgusting*. There's also a third type of concept that's both descriptive and evaluative. Philosophers call these "thick" concepts. *Cruelty* is a thick concept. When you say that an act is cruel, you're describing it and disapproving of it in a single breath. Asking yourself whether a cruel act is morally right or wrong is otiose because the moment you decided that it was cruel you thereby committed yourself to the view that it was wrong.

---

*Of course, I don't mean to say that this is *all* that there is to cruelty.

There's an important connection between evaluative judgments and motivation. When a person sincerely judges that an act is morally wrong, this entails that they want to avoid it, and that they believe everyone else should avoid it, too. We can say, very clumsily, that an act that's judged to be wrong has an element of "to-be-avoided-ness" built into it. Of course, people often do things that they think are wrong. Sometimes, temptations pulling in the opposite direction are just too strong to resist. But even in these cases, there's an inclination to avoid the morally offensive act. That's why when we do something that we believe to be wrong, we end up feeling guilty and conflicted about what we've done.

In short, moral disapproval tends to inhibit action.

Given that *wrongness* is built in to the concept of cruelty, anyone who considers an act to be cruel must be motivated not to perform it. Most of us have fantasies about treating other people cruelly. Who hasn't imagined paying someone back for a wrong they've done? Think about an occasion when you considered harming someone because you want to settle a score with them. You badly wanted to harm this person, but because you thought it would be wrong to turn your vengeful day dreams into reality, you didn't go through with it. Or maybe you went through with it, but in a less florid fashion than you fantasized (you sent them a nasty e-mail message instead of gutting them with a machete). The amount of guilt that a person experiences when contemplating a cruel act is proportional to the degree of cruelty that they judge it to possess. And the degree of guilt is proportional to the degree of inhibition. That's why acts of homicide and torture are very difficult to carry out, at least under normal circumstances. This may sound peculiar, given the catalog of horrors that I've documented in this book. But it's true. The puzzle remains that people *do* perform extravagantly cruel acts and, as Daniel Goldhagen pointed out in a passage quoted in Chapter Five, "they do it with zeal, alacrity, and self-satisfaction, even enjoyment."[24] How can we reconcile these incongruous images of the human animal?

People readily indulge in horrendous acts if they don't believe that what they're doing is cruel. There are a couple of ways that this can happen. Some people lack a moral sense (in the awful jargon of psychiatry, they have an "antisocial personality disorder"). These people are morality-blind, just like some people are color-blind. Because of this, the notion of cruelty doesn't make any sense to them. They are incapable of feeling guilt, and can do anything with a clear conscience. These people are rare. Much more often, people are able to engage in spectacularly cruel actions because they've selectively decommissioned their moral inhibitions. This is where dehumanization enters the picture. To understand how and why, we need to examine the notion of harm.

What is harm? If you harm someone, you damage them. That's clear. The damage might be physical or psychological, direct or indirect. Its medium might be a word, a silence, a glance, or the thrust of a knife into tender viscera. But harm and damage aren't identical. It's possible to do damage without doing harm, because the concept of harm applies only to certain sorts of things. Inanimate objects can be damaged, but it's impossible to harm them. When you have an automobile accident you may damage your car, but you don't harm it, and when you take it to the body shop, it's to repair the damage to it, not the harm. But if someone had been injured in the accident they would be harmed rather than just damaged (even though they might sue you for *damages*). Right now, I'm writing these words in a fifth-floor apartment in Ithaca, New York. There's a heat wave going on, so all the windows are open and there's a fan buzzing away a few feet from my chair. I'm trying to concentrate on writing this chapter, and the noise from the fan is getting on my nerves. What if I got up and hurled the fan out of the window? This would irreparably damage the fan, but it wouldn't *harm* it at all, because, obviously, fans aren't the sort of things that can suffer harm. The Chihuahua in the next apartment is making an irritating yapping noise. What if I went over there and chucked *it* out of the window, too? This would definitely harm the dog, rather than merely damaging it, because unlike fans, dogs can be harmed.

Harm is a thick concept. It's best understood as *morally unacceptable damage*. This definition implies that you can only harm a thing if it has some moral standing. There's an important connection here with pain and suffering, but not the connection that Goodall had in mind in her discussion of cruelty. Harm can be painless, and pain can be harmless, but the *kinds of beings* that can feel pain are the kinds of beings that can be harmed. Feeling pain is diagnostic of a creature's moral standing, but isn't constitutive of it.

What kinds of beings have moral standing? What makes the difference between things that can suffer harm and things that can't? In some cases the answer is clear: People have moral standing and inanimate objects don't—people can be harmed, but inanimate objects can only be damaged. But where do all the other life-forms stand? Oddly enough, our judgments about this depend in large measure on where we position them on the great chain of being. This ancient, discredited, prescientific model of the cosmos still unconsciously serves as a guideline for our moral judgments. Recall that the great chain of being classifies things both in terms of their descriptive properties and in terms of their value; it's therefore thick from top to bottom. Inanimate objects are at the bottom of the chain, and have no value in themselves. Microorganisms and plants don't fare much better, which is why even the most zealous vegans can weed their garden and wipe out untold millions of germs with disinfectants ("green" ones, of course) without suffering a single pang of guilt. Intuitions get foggier as we climb higher. Is swatting a mosquito cruel? How about stepping on a cockroach or skewering a writhing worm on a fishhook? Plunging a living lobster into boiling water, or gutting a trout for dinner? Killing a chicken? Slaughtering a lamb? Performing an abortion? Executing a criminal?

There's no fact of the matter about exactly where in this sequence damage gives way to harm, and destruction becomes cruelty, but the principle governing such judgments is both clear and embarrassingly narcissistic: the closer we judge a creature is to *us* on the hierarchy, the more inclined we are to grant it moral standing.

This principle has some resonance with David Hume's moral theory, which I briefly described in Chapter Two. Remember that Hume thought morality comes from sympathy, and that we have sympathy with others only to the extent that they resemble us. At first glance, this way of looking at things fits very nicely with the moral scheme of the great chain. Look at our attitude toward nonhuman species. We care much more about the kinds of animals that are closest to us on the chain than we do about those that are more remote (people who want desperately to protect mountain gorillas from poachers don't lose any sleep over the fate of the tiny parasites that scramble around in their fur).

Despite appearances, the two approaches are incompatible. Here's why.

Hume was what philosophers call an *empiricist*. In everyday speech, an empiricist is somebody who relies entirely on observation as a source of knowledge. Philosophers give the word a somewhat more technical meaning. In philosophy-speak, empiricism is the theory that our knowledge of the world boils down to knowledge of our sense impressions and that we "construct" our picture of the world solely from these raw materials. Look around the room. You see objects like windows and furniture. Perhaps you see other people in the room as well. Empiricists claim that all that you're *really* seeing are visual impressions— colored patches of various shapes and sizes from which you construct a picture of the room. Empiricism implies that objects (for example, the book that you are reading) are really just bundles of sense impressions. Hume's older contemporary George Berkeley, illustrated this idea using the example of a cherry.

> I see this cherry, I feel it, I taste it . . . it is therefore *real*. Take away the sensations of softness, moisture, redness, tartness, and you take away the cherry. Since it is not a being distinct from sensations; a cherry, I say, is nothing but a congeries of sensible impressions, or ideas perceived by various senses: which ideas are united into one thing (or have one name given them) by the mind.[25]

When Hume spoke of resemblance, he meant similarity in appearance—because, as an empiricist, he thought that appearances are all that we can ever know. But we've seen that appearances play second fiddle to essences in our intuitive judgments about natural kinds, and the great chain of being is a hierarchical classification of natural kinds. So it can't be that judgments about creatures' moral standing are based on their appearance. Instead, these judgments must be based on our beliefs about their essences—the kinds of things that they are. Retuning to an example used earlier in this book: if Dracula existed he would not be a creature with the same moral standing as a human being because, even though he looks human, he isn't a member of the human kind (to appreciate this, think of how odd it would be to put him on trial for his "crimes"). Our intuitive moral psychology seems to conform to the following principle: *We grant moral standing to creatures to the extent that we believe that their essence resembles our own.*

This principle points to another reason why chimpanzees can't be cruel. I've already argued that for a chimpanzee to be cruel, it would have to grasp the concept *harm*. Now it's clear that to understand harm you've got to have a notion of natural kinds. Minimally, you've got to have a notion of your own kind. So, if I'm right, a chimp could be cruel only if it could conceive of the creature that it is brutalizing as the same or a similar kind as itself. Granting this level of cognitive sophistication to chimpanzees would stretch credulity well beyond its breaking point.

This analysis explains more about how dehumanization produces moral disengagement. To dehumanize a person is to deny that they have a human essence. However, denying that a person is human is only half the story, because it's about what people aren't rather than what they are. We've seen that dehumanizers affirm the *sub*humanity of their victims, not merely their nonhumanity. To the Nazis, Jews weren't just nonhumans; they were rats in human form. And to the *genocidaires* of Rwanda, Tutsis were cockroaches.

Dehumanized people are never thought of as charming animals like butterflies and kittens. That's because dehumanizers always identify their victims with animals that motivate violence. The thinking goes something like this: Rats are vermin, and should be exterminated. So, if Jews are rats, then they should be exterminated, too. Jews *are* rats. Exterminating rats isn't cruel, because rats have no moral standing—so, exterminating Jews isn't cruel. In fact, it's *morally good* to exterminate rats because they harm human beings by spreading filth and disease—so, it's morally good to exterminate Jews.

8

AMBIVALENCE AND TRANSGRESSION

The conclusion that we must draw from all these observances is that the impulses which they express towards an enemy are not solely hostile ones. They are also manifestations of remorse, of admiration for the enemy, and of a bad conscience for having killed him. It is difficult to resist the notion that, long before a table of laws was handed down by any god, these savages were in possession of a living commandment: "Thou shalt not kill", a violation of which would not go unpunished.

—SIGMUND FREUD, *TOTEM AND TABOO*[1]

IN THE PREVIOUS CHAPTER, I argued that Mark Twain was right to say *Homo sapiens* are the only cruel animals and the only animals that go to war. In this chapter, I'm going to assess one of his other claims. Twain wrote that in between murderous military campaigns he (humankind) "washes the blood off his hands, and works for 'the universal brotherhood of man' with his mouth." The idea so bitingly expressed in these few words is that we are hypocrites about war. We slaughter our fellow human beings in war while paying lip service to the ideal of peace.

Folk-wisdom has it that actions speak louder than words, and that words are cheap. When a person says one thing but does another, it's natural to think that the act expresses their genuine commitments and the words are nothing but a coverup. Why not apply the same reasoning to the behavior of whole populations—tribes, nations, religions, maybe even the whole human race? Perhaps, then, our vaunted commitment to peace is nothing but a sham—a colossal self-deception.

There's more than a little truth to this accusation. One reason to think so is the fact that we (contemporary Americans) go to great lengths to avoid acknowledging the simple and obvious truth that war is all about killing people. Read the newspapers and listen to the speeches of our politicians. Young men and women are called to "serve their country" by going to war. When they're killed, we're told that they "gave their life for their country" (a foolish idea: soldiers' lives are taken, not given). But how often do you hear young people asked to go to war to *kill people* for their country? U.S. Army psychologist David Grossman drives home the point that the discourse of war is "full of denial."

> Most soldiers do not "kill," instead the enemy was knocked over, wasted, greased, taken out, and mopped up. The enemy's humanity is denied, and he becomes a strange beast called a Kraut, Jap, Reb, Yank, dink, slant, or slope. Even the weapons of war receive benign names—Puff the Magic Dragon, Walleye, TOW, Fat Boy and Thin Man—and the killing weapon of the individual soldier becomes a piece or a hog, and a bullet becomes a round.

Why all this dishonesty? Grossman suggests that the answer lies in our visceral horror of taking human life. "Killing is what war is all about," continues Grossman, "and killing in combat, by its very nature, causes deep wounds of pain and guilt. The deceptive language of war helps us to deny what war is really about, and in doing so it makes war more palatable."[2] If Grossman is right, then Twain's cynical assessment is too harsh. We butcher human beings and condemn killing because we're torn between two conflicting attitudes. We humans like to kill. We find it pleasant, exciting—even intoxicating. But we're also horror-struck and sickened by the spilling of human blood. Both attitudes are genuine, and both are part of human nature.

MORAL INJURY

> I thought dying for your country was the worst thing that
> could happen to you, and I don't think it is. I think kill-
> ing for your country can be a lot worse.
>
> —SENATOR BOB KERREY[3]

One of the most compelling demonstrations of the violent, demonic
aspect of human nature is the eerie elation that ordinary men sometimes experience in the heat of combat. There are many examples of
this in the literature of war. I'll quote a few to give their flavor.

We can start with Israeli military psychologist Ben Shalit, who recounts an observation that he made during his national service in the
navy.

> The gunner . . . was firing away with what I can only de-
> scribe as a beatific smile on his face. He was exhilarated by
> the squeezing of the trigger, the hammering of the gun,
> and the flight of his tracers rushing out into the dark shore.
> It struck me then (and was confirmed by him and many
> others later) that squeezing the trigger—and releasing a
> hail of bullets—gives enormous pleasure and satisfaction.
> These are the pleasures of combat, not in terms of the intel-
> lectual planning—of the tactical and strategic chess game—
> but of the primal aggression, the release, and the orgasmic
> discharge.[4]

What was the gunner experiencing? We'll never know. But other
descriptions are more explicit. J. Glenn Gray, the philosopher-soldier
whom I introduced back in Chapter One, goes further:

> Anyone who has watched men on the battlefield at work
> with artillery, or looked into the eyes of veteran killers fresh
> from slaughter, or studied the bombardiers' feelings while

smashing their targets, finds it hard to escape the conclusion that there is delight in destruction.[5]

German writer Ernst Jünger's recollection of his service in World War I gives a taste of the "combat high" from a first-person perspective.

> With a mixture of feelings, evoked by bloodthirstiness, rage, and intoxication we moved in step, ponderously but irresistibly toward the enemy lines. . . . I was boiling with a mad rage which had taken hold of me and all the others in an incomprehensible fashion. The overwhelming wish to kill gave wings to my feet. . . . The monstrous desire for annihilation, which hovered over the battlefield, thickened the brains of the men and submerged them in a red fog. We called to each other in sobs and stammered disconnected sentences. A neutral observer might have perhaps believed that we were seized by an excess of happiness.[6]

Finally, Vietnam War veteran William Broyles Jr. offers chillingly frank reflections on the euphoria of slaughter. In an essay entitled "Why Men Love War," Broyles recalls a colonel who was "a true intellectual . . . a sensitive man who kept a journal." Although this man was "far better equipped for winning hearts and minds," he was given a combat command. One night, a North Vietnamese sapper unit attacked his base (sappers are elite combat engineers). Most of his combat troops were away on an operation, so the colonel had to muster "a motley crew of cooks and clerks" who routed the attackers and killed dozens of them.

> That morning, as they were surveying what they had done and loading the dead NVA—all naked and covered with grease and mud so they could penetrate the barbed wire—on mechanical mules like so much garbage, there was a look of beatific contentment on the colonel's face that I had not seen except in charismatic churches. It was the look of a person transported into ecstasy. And I—what did I do, confronted

with this beastly scene? I smiled back, as filled with bliss as he was. That was another of the times I stood on the edge of my humanity, looked into the pit, and loved what I saw there.[7]

Off the battlefield, the pleasures of violence are savored vicariously. Public executions have always been a crowd-pleaser. In countries where this form of entertainment is unavailable there is boxing, wrestling, mixed martial arts, and various team sports that simulate warfare. And, of course, there's literature, movies, and computer games, from the *Iliad* and *Star Wars* to *Modern Warfare 2*. And then there's *war porn*. An article in *Newsweek* magazine, published in 2010, explains that when the wars in Afghanistan and Iraq broke out, both the military and individual soldiers began posting combat footage on the internet.

> But almost as soon as these images became available, civilians and soldiers alike started splicing the clips together, often adding soundtracks and spreading them across the Web. Today there are thousands of war-porn videos, and they've been viewed millions of times. Like sexual porn, they come in degrees of violence, ranging from soft-core montages of rocket-propelled grenades blowing up buildings to snuff-film-like shots of an insurgent taking a bullet to the head. And even as the U.S. begins its march toward the end of two long conflicts, these compilations continue to attract viewers. With a videogame sensibility, they fetishize—and warp—the most brutal parts of these high-tech wars.[8]

Judging from these examples, human nature is violent in the extreme. This is exactly what Richard Wrangham's hypothesis that we have inherited a dominance drive from our primate ancestors would suggest. But matters aren't so simple. When you read these examples, you were probably fascinated, maybe even titillated by them. But you were probably also sickened by them. This feeling would be more pronounced if you encountered the carnage in the flesh rather than through the pale medium of the printed word. Smelling blood and the stomach-turning stench of ruptured entrails, seeing dismembered and eviscerated human bodies, and hearing the agonized screams of the injured, is a far cry from merely reading about them.

Now, reposition yourself. Imagine that you are the perpetrator rather than an observer. Imagine that you are directly responsible for killing, maiming, and mutilating other human beings.

How do you feel?

In the movies, it's all very easy. You just pull the trigger and blow away the enemy. And death is usually tidy—there's a corpse with a barely detectable bullet hole lying in a pool of its own blood. In real life, things are different. Killing is *hard*.

One of the first people to publically acknowledge this was a controversial U.S. Army historian named Samuel Lyman Atwood ("SLAM") Marshall. Marshall got his information by talking to U.S. infantrymen immediately after firefights in the European theater of World War II, and he claimed that these conversations revealed that the majority of these men—up to three quarters of them—never fired their weapon at an enemy soldier, even when under attack. Marshall wrote about this problem (which he called the "ratio of fire") in a short but influential book entitled *Men Against Fire*. In the book he pointed out that men enter military service with a fully formed set of moral convictions, the most important of which is that it's wrong to take human life.

> He is what his home, his religion, his schooling, and the moral code and ideals of his society have made him. The Army cannot unmake him. It must reckon with the fact that he comes from a civilization in which aggression, connected with the taking of life, is prohibited and unacceptable. The teaching and ideals of that civilization are against killing, against taking advantage. The fear of aggression has been expressed in him so strongly, and absorbed by him so deeply and pervadingly—practically with his mother's milk—that it is part of the normal man's emotional makeup. This is his great handicap when he enters combat. It stays

his trigger finger even though he is hardly conscious that it is a restraint upon him. Because it is an emotional and not an intellectual handicap, it is not removable by intellectual reasoning such as: "Kill or be killed."

It is therefore reasonable to believe that the average and normally healthy individual—the man who can endure the mental and physical stresses of combat—still has such an inner and usually unrealized resistance toward killing a fellow man that he will not of his own volition take life if it is possible to turn away from that responsibility. . . . At the vital point he becomes a conscientious objector, unknowing.[9]

This is obviously a problem from a military perspective. Although it sounds very nasty, and Marshall never put it quite this way, his observations imply that military training should concentrate on overriding the recruit's moral integrity, so that he or she will have no scruples about killing on command. Moral reservations are—in Marshall's words—a "handicap" that prevents the soldier from doing his job.

Of course, these traditional methods would be out of place and ineffective in a modern military context. The U.S. armed forces overhauled their system of military training to try to solve the problems that Marshall identified. They began to train soldiers to fire immediately at man-shaped targets that pop into view, instead of the static, bull's-eye targets used during World War II and earlier. Apparently as a result, U.S. soldiers' ratio of fire increased during the Korean conflict, and by the time the Vietnam War rolled around, American troops had become much more efficient killers. But this solution created a whole new problem. The troops did better in battle, and the ratio of fire skyrocketed, but so did the incidence of combat-related psychological disorders.[10]

Marshall suggested in *Men Against Fire* that there is a strong connection between the terror of killing and what's nowadays called Post-Traumatic Stress Disorder (PTSD). He wrote, "Studies by Medical Corps psychiatrists of combat fatigue cases . . . found that fear of killing, rather than fear of being killed, was the most common cause of battle failure."[11]

This implied that killing is traumatic—so traumatic that it can precipitate psychological breakdown. Soldiers' reports of their personal experiences give this credence. Sometimes they describe becoming emotionally detached or dissociated, which makes war seem unreal, like a dream or a movie, and which insulates them from the moral enormity of their actions. Others describe shaking uncontrollably, vomiting, losing bladder and bowel control, and being overwhelmed by feelings of guilt. Historian William Manchester's description of killing a Japanese sniper on Okinawa is a good example of how the experience of taking life in close combat can impact on the killer. The sniper had been firing on Manchester's unit. He saw that the shots were coming from a fisherman's shack, and decided to enter it. He kicked the door open and caught the man off guard. Manchester's first shot was wide of the mark, but the second one hit the sniper in the heart. "He dipped a hand in it and listlessly smeared his cheek red. . . . Almost immediately a fly landed on his left eyeball," as the frightened young Marine pumped bullet after bullet into the slumping corpse. He then paused.

> I don't know how long I stood there staring. A feeling of disgust and self-hatred clotted darkly in my throat, gagging me. Jerking my head to shake off the stupor, I slipped a new fully loaded magazine into the butt of my .45. Then I began to tremble, and next to shake, all over. I sobbed, in a voice still grainy with fear: "I'm sorry." Then I threw up all over myself. I recognized the half-digested C-ration beans dribbling down my front, smelled the vomit above the cordite. At the same time I noticed another odor; I had urinated in my skivvies. . . . I knew I had become a thing of tears and twitching and dirtied pants. I remember wondering dumbly: Is that what they mean by conspicuous gallantry?[12]

Observations of a connection between combat guilt and psychological damage go back a long way. In his sensitive study of psychological trauma suffered by veterans of the American Civil War, Eric T. Dean notes that these men sometimes felt that they had been tainted by an unpardonable sin.

> For instance, one veteran was operating under the delusion that he had been accused of murder and that a corpse had been secreted in his house. Another thought that he was guilty of heinous crimes committed during his early life. Others were brooding over transgressions or convinced that they were hopeless sinners: "said he was guilty of great crimes . . . he thinks he is lost for all eternity"; "delusion seems to be that he has done something terrible."

Later, during World War I, Nobel laureate Jane Addams described "hearing from hospital nurses who said that delirious soldiers are again and again possessed by the same hallucination—pulling their bayonets out of the bodies of men they have killed."[13]

Rachel MacNair points out in her book *Perpetration-Induced Traumatic Stress* that even Nazi killers, who are conventionally portrayed as monsters devoid of even a shred of moral sensibility, had difficulty stomaching the work of extermination. For example, Rudolf Höss, the first commander of Auschwitz, reported that Adolf Eichmann had told him, "Many of the *Einsatzkommandos*, unable to endure wading through blood any longer, had committed suicide. Some had even gone mad. Most of the members of the *Kommandos* had to rely on alcohol when carrying out their horrible work."[14]

Psychological studies have strongly confirmed the relationship between killing and psychological damage. A study of almost 3,000 U.S. Army soldiers by University of California psychologist Shira Maguen and her coworkers found that the 40 percent of them who reported killing in combat were significantly more prone to psychological problems than the rest. This effect was independent of combat exposure—in other words, it can't be explained by saying that the men who killed in combat were also the men who were most extensively engaged in combat and therefore exposed to other stressors. In their 2009 report, they demonstrated "highly significant" correlations between killing in combat and the severity of PTSD, dissociation, violent behavior, and general psychological impairment. And then, of course, there's suicide. In the United States, veterans are twice as likely as members of the general population to die from suicide, and research suggests that veterans' suicides are often linked to persistent feelings of guilt about what they've done. A publication by the United States Department of Veterans Affairs plainly states, "Research suggests that for veterans, the strongest link to both suicide attempts and thinking about suicide is guilt related to combat. Many veterans have very disturbing thoughts and extreme guilt about actions taken during times of war."[15]

The label "PTSD" presents the psychological effects of combat as a "disorder." But is it? Is a bullet wound or the loss of a limb a disorder? Psychiatrist Jonathan Shay doesn't think so. He argues that psychological damage is an injury, and coined the term *moral injury*. Boston University psychiatrist Brett T. Litz took up the term, and defined it as psychological damage caused by "perpetrating, failing to prevent, bearing witness to, or learning about acts that transgress deeply held moral beliefs and expectations."[16]

An article by Litz and five coauthors published in 2009 in the *Clinical Psychology Review* notes that there's considerable evidence that moral injury has powerful negative psychological consequences. They give some sobering statistics. In 2003, 32 percent of U.S. marines and soldiers reported that they were responsible for the death of an enemy combatant, and 20 percent admitted responsibility for the death of a noncombatant. Perhaps even more significantly, 27 percent reported that they faced ethical challenges in combat to which they didn't know how to respond.* Almost a third of U.S. combat troops experienced significant moral conflict during their deployment, and this figure very likely underreports the true extent of the problem. Small wonder, then, that of the approximately 1.7 million military personnel that have served

---

*Interestingly, less than half of the men and women surveyed thought that noncombatants should be treated with "dignity and respect" and about a third admitted to having cursed at or insulted them.

in Iraq and Afghanistan, as many as three hundred thousand—close to 17 percent—may be suffering from PTSD and many more may be suffering from less easily diagnosable psychological injuries.[17] Litz and his coauthors note that:

> We are doing a disservice to our service members and veterans if we fail to conceptualize and address the lasting psychological, biological, spiritual, behavioral, and social impact of perpetrating, failing to prevent, or bearing witness to acts that transgress deeply held moral beliefs and expectations, that is, moral injury.[18]

Why does killing produce such guilt? Marshall reasoned that it comes from social programming. Taking human life is the ultimate forbidden act. It is interdicted by a social taboo that runs so deep that it can't easily be sloughed off.

If this explanation is correct, then soldiers from cultures with a more permissive attitude toward bloodshed should function more effectively in combat. In the developed world, killing is only permissible at the behest of the state, but in some traditional cultures individuals have much greater latitude. There are societies in which men are expected to avenge the death of members of their family or clan by taking a life in return. If this happened in Los Angeles, we'd think of it as a gangland murder. The cultural context makes all the difference.

In a controversial article published in *The New Yorker* in 2008, Jared Diamond claims that children growing up in such societies are exposed to bloodshed at an early age, and suggests that this early conditioning makes them guilt-free killers in adulthood.

> Traditional New Guineans . . . have from childhood onward often seen warriors going out and coming back from fighting; they have seen the bodies of relatives killed by the enemy, listened to stories of killing, heard fighting talked about as the highest ideal, and witnessed successful warriors talking proudly about their killings and being praised for

them. If New Guineans end up feeling unconflicted about killing the enemy, it's because they have had no contrary message to unlearn.[19]

Diamond contrasts this allegedly easygoing attitude with that of American veterans who so often return from war conflicted about what they've done. "It's no wonder," he remarks, "that many soldiers who kill suffer post-traumatic stress disorder. When they come home, far from boasting about killing, as a Nipa tribesman would, they have nightmares and never talk about it at all, unless to other veterans." This comparison is misleading on two counts: it incorrectly equates feuding and warfare, and it presents a shallow, two-dimensional picture of the psychology of men and women in traditional societies. Feuding and war both aim at killing, but there's a profound difference between them. In a feud, the killing is personal. The killers have a score to settle, and this is a powerful motive for extracting revenge. In contrast, war is impersonal—hostilities are directed at an abstraction, "the enemy," rather than at any individual. Feuders see themselves as putting something right, whereas soldiers see themselves as doing their duty.[20]

Soldiers often lose their nerve precisely when they're confronted with the fact that those whom they are trying to kill are fellow human beings who they have no grudge against. This principle is famously illustrated by a passage written by George Orwell describing an experience that he had during the Spanish Civil War. Early one morning, Orwell and a comrade set out to search for fascists. Hiding in a ditch to avoid being spotted by the enemy, they watched as:

> At this moment, a man presumably carrying a message to an officer, jumped out of the trench and ran along the top of the parapet in full view. He was half-dressed and was holding up his trousers with both hands as he ran. I refrained from shooting at *him*. It is true that I am a poor shot and unlikely to hit a running man at a hundred yards. . . . Still, I did not shoot partly because of that detail about the trousers. I had come here to shoot at "Fascists"; but a man who is

holding up his trousers isn't a "Fascist," he is visibly a fellow-creature, similar to yourself, and you don't feel like shooting at him.[21]

Lieutenant Emilio Lussu, an Italian soldier who fought in World War I, recounts something similar. During the night, Lussu had crept into a position overlooking the Austrian trenches. As dawn broke, he could plainly see them "as they really were, men and soldiers like us, moving about, talking, and drinking coffee." Lussu spotted a young officer, and took aim. At that moment, the Austrian lit a cigarette. "That cigarette formed an invisible link between us," he wrote. Lussu knew that it was his duty to shoot, but

> I began to think that perhaps I ought not to do so. I reasoned like this: To lead a hundred, even a thousand, men against another hundred, or thousand was one thing; but to detach one man from the rest and say to him, as it were, "Don't move. I'm going to shoot you. I'm going to kill you"—that was different. . . . To fight is one thing, but to kill a man is another. And to kill him like that is to murder him.[22]

William D. Ketcham, a Union veteran of the American Civil War, told a story of aiming at a Confederate officer, pulling the trigger, and missing his mark. "I did not elevate the sight," he later confessed. On another occasion, Ketcham shot a man and afterward, when he inspected the corpse, noticed that it had multiple gunshot wounds. "[T]hat survey of the target satisfied my mind that I was not responsible for his death," he wrote in his memoir, "and his blood was not on my hands and I have always been glad that I knew that fact."[23] Because blood feuds are motivated by passion, it's easy to see the killers would be less inhibited—less troubled by moral scruple—than soldiers are. Imagine that you found your parent or child, brother or sister, lying face down in the dirt with an arrow protruding from their back. Wouldn't you be capable of doing almost anything to avenge them?

In war, the motivation to kill is nowhere near as strong, because the soldier kills out of duty rather than out of passion. That's why wars require propaganda to drum up motivation. And one of the most popular themes in propaganda is to represent the war as a feud. To personalize it by inducing potential combatants to believe that their families are threatened by the enemy, who wants nothing better than to kill their mother, rape their sister, and bayonet their baby.

Even though feuding tribesmen have a stronger motivation to kill than soldiers normally do, it would be wrong to assume—as Diamond seems to think—that they don't have any inhibitions at all. To suppose this is a modern version of the old, ethnocentric fantasy of the bloodthirsty savage. If you want to really understand a culture's attitude toward lethal violence, you've got to look beyond the superficial braggadocio of warriors. Irenäus Eibl-Eibesfeldt, whose work I briefly discussed in Chapter Two, points out that tribal people are typically ambivalent about war. For instance, he found that in mourning ceremonies for fallen warriors among the Melpa of Papua New Guinea, "war was characterized as evil and associated with guilt," noting as well that this "does not contradict the fact that one can participate in it with enthusiasm, and a certain athletic zeal, for guilt and enthusiasm can be activated simultaneously."

> It is probably for this reason that the attacker, to excuse himself, typically claims that the other parties initiated the hostilities, and that he was compelled to protect himself. This is a common position taken by members of traditional societies and representatives of civilized nations, regardless of the type of government.[24]

Ceremonial purification after battle is another manifestation of ambivalence about killing. In many cultures (including the fearsome Yanomamö), warriors who have killed are required to undergo ritual purification before reentering society. Freud discussed this in his book *Totem and Taboo*, where he interpreted cleansing rituals as manifestations of guilt. The anthropologist Harry Holbert Turney-High followed Freud's example in his book *Primitive War*:

War and killing push men into some kind of marginality
which is at least uncomfortable, for there seems to be a basic
fear of blood-contamination, an essential dread of human
murder. If man did not consider human killing something
out of the ordinary, why has there been such common fear of
the enemy dead, the idea of contamination of even a pres-
tigeful warrior of the we-group? We have seen that the chan-
neling of frustration into hatred toward the enemy is good
for the internal harmony of the we-group, but the enemy is
human, too. Humanity is capable of ambivalent attitudes towards its enemies.[25]

There's another reason to be skeptical of Marshall-style explanations. If the reluctance to kill is purely the result of learning to conform
to a social taboo, why is this taboo so much more powerful than the
other ones? Normally, we don't have a lot of difficulty violating social
rules. For example, we're not supposed to lie, but most of us lie a great
deal; we're not supposed to steal, but every year many people evade paying income tax; and we're not supposed to commit adultery, but infidel-
ity is all too common. All of these social taboos forbid us to do things
that we're strongly tempted to do, which is why they're often more hon-
ored in the breach than in the observance. However, the interdiction
against killing doesn't seem to fit this pattern at all. How many are unable to resist a temptation to kill others? Homicide is remarkably rare,
even in countries with elevated homicide rates. In Colombia, South
America, which consistently has one of the highest homicide rates in
the world, there were on average only 62.7 killings per 100,000 people
between 2000 and 2004. Why is killing so different from lying, stealing,
and adultery? When we consider people's learning histories, the dispar-
ity is even starker. Most people have never been told not to kill another
human being, nor have they been punished for killing or threatening to
kill someone. But they've frequently been told by not to lie, steal, and so
on, and have been punished for these infractions. One would imagine
that, if resistance to killing boils down to social learning, killing should
be a lot easier to do than lying and stealing. But it isn't.

THE INFORMATION EXPLOSION

Humans may be hard-wired to get edgy around the
Other, but our views on who falls into that category are
decidedly malleable.

—ROBERT SAPOLSKY, "PEACEFUL PRIMATES"
A NATURAL HISTORY OF PEACE[26]

If our resistance to killing isn't a result of learning all on its own, then it
must be based in part on something innate. There must have been some
feature of our evolution that accounts for why humans have such robust
inhibitions against taking human life. Unfortunately, it's all too easy to
dream up evolutionary scenarios about how this might have happened.
Facts about the social behavior of our remote ancestors are thin on the
ground, and it isn't helpful to churn out hypotheses that have no evidence to answer to. But it's not helpful to throw in the towel either. So,
I'm going to tread a cautious middle course. I'm going to set out what I
think is a plausible evolutionary hypothesis of how dehumanization
became part of our psychological repertoire. I don't claim that it's the
best possible explanation. There may well be better ones. And I'm
certainly not claiming that it's true. But it's the best story that I've
been able to come up with so far, and I think that it's worthy of con-
sideration.

In the beginning was the ape. As I've mentioned, around six and a
half million years ago our ancestors were similar to chimpanzees.
They had brains roughly the same size as chimpanzee's brains, and
probably had roughly similar mental abilities. It is reasonable to think
that the social behavior of this ancestral ape was similar to that of
chimpanzees—i.e., that they lived in fission-fusion communities,
were territorial, xenophobic, and didn't hesitate to kill their neighbors.

For the next four million years, evolution of the human lineage proceeded slowly. Then, about two million years ago, a new
primate species called *Homo erectus* stepped out on the African

savannah.* *Erectus* looked much more human than any of its prede-cessors, but what was most remarkable about the new creature was its brain, which doubled in volume over the next million years. *Homo erectus* got smart—much smarter than any primate species that came before. They created large stone tools (the beautifully crafted Acheu-lean "handaxes") and built campfires. They probably hunted coopera-tively and consequently consumed more meat than their predecessors. They may even have invented cooking.[27] *Homo heidelbergensis* ("Hei-delberg Man") evolved from *Homo erectus* approximately 600,000 years ago, and *Homo sapiens* evolved from *Homo heidelbergensis* roughly 200,000 years ago. Soon after, anatomically modern humans— *Homo sapiens sapiens*—came into being. These people were almost exactly like you and me physically, but they were very different men-tally, in ways that I'm going to clarify in just a moment.

The forces that transformed anatomically modern humans into psy-chologically modern humans were mainly cultural. Human culture was born in two great spasms. The first began about forty thousand years ago in what's known as the Upper Paleolithic revolution. As prehistoric humans transitioned into this period, stone tools made a huge leap in sophistication. They became more specialized and incor-porated new materials such as bone and antler. For the first time, peo-ple hafted stone points to wood to make spears and harpoons, and used spear-throwers to increase their force and distance. Soon works of art appeared—beautiful paintings on cave walls across southern Europe, and small stone and ivory sculptures. Then, eight to ten thousand years ago, the Neolithic revolution exploded in the Middle East (and some-what later elsewhere) as human societies underwent a cascading se-quence of transformations. People abandoned an economy based on hunting and gathering, and turned to agriculture and animal hus-bandry. They domesticated wild plants and animals, and anchored themselves in permanent settlements. Agriculture led naturally to the

---

*Sometimes this hominin is called *Homo ergaster* rather than *Homo erectus*. There is a great deal of controversy about the relationship between the two species (if, indeed, they are two species).

ownership of land, and to technologies for cultivation, irrigation, and storing surplus grain. As settlements expanded into towns and cities, increased population density made way for social stratification, divisions of labor, and increased trade. The new towns and cities required administrative structures, which were interwoven with religious institutions.

During their journey through time, prehistoric humans acquired a range of mental aptitudes that we take for granted today, but which were almost certainly absent during the earliest stages of hominid evolution. These include the psychological traits that are necessary conditions for dehumanization, namely:

1. A domain-specific *folk-biology module* responsible for parsing the biological world into natural kinds (species) and making inferences about them.
2. A domain-specific *folk-sociology module* responsible for parsing the social world into natural kinds (ethnoraces) and making inferences about them.
3. A domain-general *capacity for second-order thought* that makes it possible to reflect on one's own mental states.
4. An intuitive *theory of essences* used to explain why there are natural kinds.
5. An intuitive *theory of natural hierarchy* (great chain of being) for ordering the natural world.

It's reasonable to think that our common ancestor with the chimpanzee didn't have any of these (or at best had only the first of them), and we know that present-day humans have all of them. So, all five psychological traits must have emerged over the last six and a half million years or so. Determining when all five of these pieces were in place should allow us to set a lower limit for when dehumanization began. Of course, this sort of reasoning won't tell us exactly when people started dehumanizing each other, because although the five conditions that I listed above are individually necessary, they may not be (in fact, they probably aren't) jointly sufficient.

Because the evidence is indirect and often sketchy, it's difficult to draw conclusions about prehistoric people's psychology. But it's possible to make some educated guesses, and that's what I'm going to do right now.

We can begin with the folk-biology module, which is probably the most ancient of the five. It's very helpful for an animal to be able to distinguish what it eats from what eats it, and any animal that does this has to have at least a rudimentary ability for responding differentially to different biological kinds. But as we saw in the example of ant "warfare," an animal can behave differently in relation to different species without having concepts of biological kinds or being able to make inferences about them. The predator-detecting abilities of fancier animals—for example, vervet monkeys—are closer to what we're looking for. Vervet monkeys distinguish between leopards, pythons, and eagles, and give distinctive alarm calls for each, but it's not clear that they have concepts of these animals.[28] Chimpanzees probably have primitive concepts of biological kinds, and our common ancestor may have had them as well. However, we can be certain that a well-developed folk biology was in place by the time our prehistoric ancestors had become accomplished hunters. People who depend on hunting for a significant portion of their diet need to have detailed knowledge of the animals that they hunt (this is also true, to a more limited extent, of the intellectual demands of gathering). They need to have a refined understanding of the way that a variety of animals behave under a range of circumstances, They need to have expert tracking skills, including the ability to identify hoof-and pawprints, various kinds of feces, and other marks that animals leave on their environments, and when they migrate into new territories, they have to be able to quickly learn about the behavior of newly encountered species—in part, by making similarity-based inferences.

Archaeological evidence suggests that meat was a significant component of *Homo erectus's* diet, although opinion is divided about whether this came primarily from hunting or from scavenging the carcasses of animals brought down by large predators. Whatever the facts are about *Homo erectus*, we know that prehistoric *Homo sapiens* excelled at cooperative hunting. We also know that they dispersed out of Africa to exploit a wide range of habitats where they encountered, and learned to hunt, many unfamiliar species. None of this would be possible unless they were competent to make inferences about animal behavior. So a well-developed module for intuitive folk-biology is likely to have been in place by the middle Paleolithic period, perhaps 100,000 years ago.[29]

The next puzzle concerns our knack for second-order thought. Many cognitive scientists believe that second-order thought is tightly bound up with language. The idea is that being able to use and understand language makes it possible for a person to reflect on their own mental states. Philosopher and cognitive scientist Andy Clark explains how this works:

> Rather amazingly, we are animals who can think about any aspect of our own thinking and can thus devise cognitive strategies . . . aimed to modify, alter, or control aspects of our own psychology. . . . [A]s soon as we formulate a thought in words . . . it becomes an object both for ourselves and for others. As an object, it is the kind of thing that we can have thoughts about. In creating the object, we need have no thoughts about thoughts, but once it is there, the opportunity immediately exists to attend to it as an object in its own right. The process of linguistic formulation thus creates the stable structure to which subsequent thinkings attach.[30]

These don't have to be words that come out of one's mouth. They can be words that one silently "says" in one's head. The point is that once a person expresses a thought as a sentence, in whatever medium, then he or she can think about that sentence. So, once prehistoric people were able to do this, they were in a position to think about their own thoughts. What a remarkable achievement this was! Tufts University philosopher Daniel Dennett hypothesizes that higher-order thought emerged in two steps. Early on, hominid brains had various domain-specific cognitive systems. These special-purpose mental organs weren't integrated with one another—there wasn't any information flow between them. But even though these cognitive systems operated autonomously, they were adequate for negotiating the practical challenges that these primates had to face—challenges like obtaining food, avoiding predators, finding mates, and so on. Then language evolved. Linguistic communication was a tremendous boon to those who could use it. Valuable information could rapidly move from mouth to ear, and thereby hop from brain to brain, with high fidelity. This vastly accelerated the speed of cultural transmission.

Without language to help it along, the spread of ideas is painfully slow. You can get a sense of just how slow from observations of cultural transmission among nonhuman primates. One of the most impressive examples concerns a community of Japanese macaques living on the island of Koshima. In 1953, scientists studying these monkeys began provisioning them with pieces of sweet potato. The potatoes were placed on the beach, so to avoid getting their mouths full of grit, the macaques had to get rid of the sand clinging to the food before eating it, which they accomplished as best they could by brushing off the sand with their hands. Then one day a young female named Imo (Japanese for "yam") discovered that she could rinse the potatoes in river water, which noticeably produced a superior result. Before long other macaques (first her siblings, and then her mother) began to wash their potatoes, too, and the practice eventually spread to the entire troop. Today, over fifty years later, descendents of the original group continue to wash their potatoes (nowadays in salt water, perhaps as a method of seasoning).[31] But even among clever Japanese macaques it took *nearly a decade* for potato washing to become a cultural fixture.

Language turbocharged the spread of culture, but communicating ideas wasn't the only thing that language was good for. It was also good for thinking with. Language reconfigured the way that human brains process information by bringing second-order cognition into being. Here's Dennett's story of how this happened:

> Then one fine day (in this rational reconstruction), one of these hominids "mistakenly" asked for help when there was no helpful audience within earshot—except itself! When it

heard its own request, the stimulation provoked just the sort of other-helping utterance production that the request from another would have caused. And to the creature's delight, it found that it had just provoked itself into answering the question.[32]

Of course, Dennett doesn't want us to take his charming fable literally. His point is that once language became established, people could use it to elicit information from themselves that they weren't aware that they had. Language did more than enable people to talk to other people; it also permitted parts of brains to "talk" to one another through the medium of linguistic thought. Once this happened, the mind became what Dennett calls a *Joycean machine*: a site of ongoing inner dialogue pulling together information from far-flung regions of the brain.

This is a good story, but is there any evidence that things actually happened this way?

Because evidence about the psychology of long-dead people is always meager, it's easy to dismiss any attempt to reconstruct the evolution of the human mind as the wildest of speculation. But this would be too hasty. Our knowledge of the remote past comes primarily from the evidence uncovered by archaeologists. Nowadays, archaeology has a subdiscipline called *cognitive archaeology* that uses material evidence from the prehistoric past to generate and test hypotheses about our early ancestors' minds. Steven Mithen, a cognitive archaeologist at the University of Reading in the United Kingdom, argues that the archaeological evidence supports a scenario very much like the one described by Dennett. Mithen argues that our early ancestors evolved a suite of cognitive modules that equipped them for dealing with the practical exigencies of life. Then, around 200,000 years ago *Homo sapiens* evolved, and so did language. It took another 50,000 years or so for language to develop and proliferate enough to become a regular feature of human life. Once established, the use of language made it possible to integrate information from separate cognitive domains, and achieve what Mithen calls *cognitive fluidity*. This creative crisscrossing of cognitive domains heralded the explosion of culture during the Upper Paleolithic period.

From that point onward, culture had an overwhelmingly powerful impact on the development of our species.[33]

If Dennett and Mithen are correct, the evolution of language had three important consequences: it facilitated communication and therefore the spread of culture, it unified the mind by integrating domain-specific modules, and it made second-order thinking possible. Now, here's an interesting question. What did the original second-order thinkers think about? What reflections preoccupied them the most? This question isn't quite as baffling as it might seem, because these women and men left a record of their thoughts in their works of art. The magnificent murals that adorn the walls and ceilings of caves across southern Europe are festooned with realistic portraits of animals: horses, deer, cattle, bison, lions, birds, bears, goats, mammoth, and wooly rhinoceros. There's practically *nothing else* in these pictures—no trees, clouds, mountains, rivers, campfires, or shelters—and the same is true of their sculptures and engravings.

Although painting a picture or carving a sculpture is, in many ways, different from putting a thought into words, the two forms of expression have something important in common. When you put a thought into words, or an idea into a picture, you've externalized it. You've turned your thought into a representation: a thing that you can think about. But unlike spoken language, which is transient, art leaves a record. So, by looking at people's art, we can get insights into what sorts of things they reflected on.

Paleolithic art shows that these men and women were deeply—one might even say, obsessively—preoccupied with thoughts about biological kinds. They were reflecting on their thoughts about animals, and asking questions about those thoughts. Prior to this point, prehistoric people had biological concepts—for example, the concept *porcupine*—but now, for the first time, they could wonder about what it is that makes something a porcupine. It seems reasonable to think that they came to the same intuitive conclusion that people do today. They concluded that what makes an animal an animal of a certain kind is its essence. Porcupines are porcupines because they have the porcupine essence. The members of every species have something inside of them that makes them members of that species.

There are also representations of human beings in Paleolithic art. The human presence in art shows that men and women of the Upper Paleolithic weren't just thinking about horses and bison. They were also thinking about themselves. Following the same pattern of explanation, we can suppose that, when Paleolithic people asked themselves, "What makes us human?" the answer that they gave was, "Having a human essence."

Let's suppose that this is what happened. Conceiving of human beings as having a shared essence must have radically transformed their pattern of social interaction. Let me use an example to illustrate why. Imagine that you encounter a virulently racist man—a man who hates and despises black people and advocates laws that discriminate against them. If you wanted to persuade this man to adopt a more enlightened attitude, how would you go about it? I think that you'd likely appeal to a common humanity. You would call his attention to the fact that black people are human beings, just like he is. From a strictly logical point of view this doesn't make much sense. Why should a reminder of shared species membership have any impact on a person's attitudes? I think that the best way to make sense of this is as follows: Thinking of a person as a member of the same species as yourself, as sharing the same essence, automatically evokes a sense of oneness with them. You perceive them as a fellow member of the human community. By conceiving of a person in this way, you conceive of them as a member of your in-group, and this triggers inhibitions against harming them.

If I'm right, this implies that when Upper Paleolithic people began thinking of themselves as sharing an essence, this must have counteracted their xenophobic tendencies. It must have made them more willing to interact with people outside of the limiting boundaries of their communities. There's evidence that this occurred: The first unequivocal evidence of trade comes from this period, and trade can only happen if communities are open to friendly contact with one another.

> Seashells from the Mediterranean appear at Upper Paleolithic sites several hundred kilometers north in central Europe; fossilized amber from the Black Sea is found in central Russia (up to 700km away). . . . But the most compelling

examples are associated with the production of tools from certain types of stones that were routinely transported 100– 200 km (and up to 400 km for more distinctive high-quality flint) from Upper Paleolithic quarry sites in north central and eastern Europe . . . and . . . it is safe to assume that trade in purely local goods and services (e.g., perishable foodstuff, housing, and, especially, labor services) had already reached a considerable level of intensity.[34]

Increased contact between tribes encouraged people to adopt markers to signal their ethnic affiliation. Archaeological deposits dating from the Upper Paleolithic onward contain beads and pendants made from shells, animal teeth, ivory, and ostrich eggshells, and carved figurines show hairstyles and body decoration (perhaps tattooing).[35] University of Arizona archaeologists Steven L. Kuhn and Mary C. Stiner observe that this "implies an expansion in the scales of human interaction," and that people were "finding it necessary and advantageous to broadcast their identities to larger numbers of people spread across a more complex network of groups."

Increasing populations . . . changed the social landscape, putting nearly everyone in more frequent contact with strangers. This heightened level of interaction fostered heightened sensitivity to group boundaries as a means of delimiting and defending territories. . . . In an ever more complicated social landscape, there are many advantages to communicating one's identity effectively and to as many other people as possible. Such conditions in turn encouraged the development of novel modes of communicating social information, including body ornamentation. Thus began the first stages of the information revolution.[36]

Do you remember Francisco Gil-White's theory of the origin of folksociology that I described in Chapter Six? Gil-White hypothesized that, as tribal groups began consolidating ethnic identities and adopted ethnic markers, this affected how they conceived of one another. Tribes now "looked" like biological species to the human brain, and people began thinking of tribal groups essentialistically. So, if Gil-White's theory is correct, it looks like intuitive folk-sociology may have first got going during the Upper Paleolithic revolution.

Let's pause to recap the story before adding the final pieces. It begins six and a half million years ago with a chimpanzee-like progenitor—a violently xenophobic ape. This primate handed down its violent propensities to its descendents, including *Homo sapiens*. Thanks to the evolution of language, *Homo sapiens* became capable of second-order thought, and for the first time could wonder about what makes humans human. This led to the idea of a human essence that all people share. The idea that all people share an essence softened the line drawn between in-group and out-group. People began to develop friendly relations with other communities. This led to the invention of trade, which further accelerated the spread of culture. As population density increased, and contact between cultures became more and more frequent, tribes adopted ethnic markers—distinctive forms of dress, behavior, and adornment—to signal their ethnic identity. Finally, this led to the notion of ethnoraces—essentialized human groups—as the folk-biology module began to respond to ethnic groups as though they were biological species.

This all sounds very nice, but there was a worm lurking in the apple. The dominance drive inherited from our primate progenitors didn't simply vanish. Our Stone Age ancestors still had a deeply rooted tendency to treat outsiders with hostility, and to kill them when the opportunity arose. We know this from images of carnage in prehistoric art. Archaeologist Jean Guilaine and paleopathologist Jean Zammit explain that:

> A loose pebble from the Paglicci cave in the southeast of Italy was found to have been engraved with . . . a human-like figure which has been struck by several spears from the head down to the pelvis. . . . In Cougnac (Lot, France), a decapitated body is shown, struck in the back by three

projectiles, whilst another individual has been struck by seven spears all over his body. In the Pech-Merle cave in Cabrerets (Lot, France), one individual is shown having been hit by arrows all over his body, both from the front and from behind. In Combel, part of the same network of caves, a human-like figure with an animal-shaped posterior . . . can be seen collapsing, after having sustained several injuries. A carving upon a bone from Gourdon (France), showing only the pelvis and legs of a human figure . . . shows several arrows penetrating the victim's legs and rear. Also of interest is a rock engraving discovered in the cave at Sous-Grand-Lac (France). . . . The engraving shows a figure injured in both the neck and back by a number of projectiles. Arrows appear to have struck this individual's posterior and penis.[37]

This is how our ambivalence toward violence began. The new fellow feeling born of a sense of a common humanity took its place alongside the older xenophobic sensibility. On one hand, we are disposed to carve the world into *them* and *us* and take a hostile stance toward outsiders. On the other hand, we think of all people as members of the human community and have a powerful aversion to harming them. Dehumanization offered an escape from this bind. By a feat of mental prestidigitation we discovered a method for counteracting inhibitions against lethal violence by excluding our victims from the human community.

But there is still an ingredient missing from the mix. Dehumanization can't occur without the concept of subhumanity, and it's not clear that Paleolithic people had any such notion.

There are some clues in Paleolithic art. Nonhuman animals are painted and carved naturalistically and with an exquisite attention to detail, while humans are usually portrayed in a highly stylized way. This stylistic contrast suggests that the people of the Upper Paleolithic set humans apart from other animals, but this tells us nothing about the relative *value* that they attributed to humans and nonhuman animals. Perhaps noting differences between how humans and animals were treated will give us something more to work with. Animals were routinely hunted, slaughtered, and eaten. Their pelts were used to make clothing, their hides were processed to make leather, and their bones and teeth were used for ornaments. Human beings weren't treated in this way. Although cannibalism may have been practiced during the Upper Paleolithic, it wasn't routine, and artifacts made from human bones are uncommon. Humans ceremonially buried their dead rather than leaving their carcasses to rot. These facts suggest that prehistoric people considered humans to have greater moral value than nonhuman creatures.

It looks like human beings started to dehumanize one another at some point during or after the Upper Paleolithic period. This may have been quite recent, as the earliest unequivocal examples of dehumanization date from the second millennium BCE. If I am right, dehumanization caught on because it offered a means by which humans could overcome moral restraints against acts of violence. Because the folk-sociological thinking was already in place, ethnic groups were conceived as pseudospecies, each of which was imagined to have a unique essence that distinguished it from all the others. It was but a short step to imagine that some of these pseudospecies possessed a subhuman essence. This made members of the group seem like subhuman animals and therefore legitimate targets of violence. By selectively dehumanizing other communities, humans found a way to get around their ambivalence. They could selectively exclude ethnic groups from the charmed circle of a common humanity, slaughter them, and take their possessions, while at the same time enjoying the benefits of trade and affiliation with others.

PREY, PREDATORS, AND UNCLEAN THINGS

I mentioned near the end of Chapter Seven that nobody dehumanizes others by imagining them to be appealing animals. The animal has got to be one that elicits an aggressive response. In earlier chapters, I hinted that dehumanized people are often perceived either as predators or parasites. In the remainder of this chapter, I'm going to discuss the phenomenology of the main forms of dehumanization, and provide a few examples of each to underscore some of their salient characteristics.

Sometimes dehumanized people are thought to be a despised or hated "animal" of no determinate kind. However, they are more often represented as any of three kinds of creature: dangerous predators, unclean animals, or prey. There are occasional departures from this pattern, but for the most part, it is surprisingly robust across both time and place.

Let's start with unclean animals—vermin, disease organisms, and parasites. If you are like most people the sight of a bowl seething with maggots is stomach-turning. The reaction of disgust is accompanied by a peculiar sense of threat. The fear isn't that the animal itself can inflict harm—the fear of maggots isn't like the fear of poisonous snakes or snarling dogs. Rather, it's the fear that they can *contaminate* one with something harmful. That's why we are repelled by the prospect of "unclean" animals touching us, or even coming into contact with objects that we touch. Sometimes it feels like the mere sight of them can pollute us, as though their filth could enter our bodies through our eyes.

Disgust appears to be a uniquely human trait. Other animals reject food that they do not like, but they don't show signs of revulsion like humans do. Some think that this is because things that elicit disgust—things like bodily fluids, rotting carrion, and a variety of animals—do so because they are unpleasant reminders of our animal nature, but it seems more plausible that humans alone experience disgust because humans are the only animals able to reflect on their distasteful experiences. Coming into contact with something nasty is one thing—thinking of it *as* nasty is quite another.

People have an intuitive theory of contamination. We not only conceive of certain things as revolting, we also attribute their foulness to pollutants that they contain—pollutants that can get inside us and damage or even kill us if we come into contact with them. Although the propensity for disgust is innate, culture plays a huge part in determining what sorts of things elicit it. Even though you would probably find it extremely difficult to eat a piece of food crawling with maggots, until quite recently, both Europeans and Americans savored maggoty cheese. The English writer Daniel Defoe, who lived in the seventeenth and eighteenth centuries, recorded that a Stilton cheese from Huntingdon-shire was brought to the table "with mites and maggots round it, so thick, that they bring the spoon with them for you to eat the mites with as you do the cheese."[38] The practice was still popular in 1940, when Yale zoologist Alpheus Hyatt Verrill was amused to observe that:

> Grasshoppers, crickets, grubs, caterpillars, are all eaten with gusto by some races, and although few of our people could be induced to much as taste such viands, yet we eat maggoty cheese and pay fancy prices for it. But the very same persons who like cheese fairly alive with "skippers"* would be nause-ated at the thought of swallowing an apple worm, a corn-borer caterpillar, or eating a wormy chestnut or weevil-infested cereal.[39]

Likewise, animals regarded in one culture as unclean may be seen very differently at other times and places. Take dogs. To contemporary Americans and Europeans, dogs are man's best friend, but in some parts of the world they're regarded as irredeemably disgusting. Dog ownership has been denounced as "depraved" by clerics in present-day Iran.[40] The Bible contains a number of disparaging references to dogs. For example, Paul warns the Christian community in Philippi to "watch out for those dogs, those men who do evil, those mutilators of the flesh."

Dehumanized people are often seen as dangerous, unclean animals: creatures like rats, worms, lice, maggots, dogs, and bacilli. They evoke a feeling of horror (disgust mixed with fear), and arouse the urge to *ex-terminate* the offending creature. This form of dehumanization is often specifically linked to genocide. Recall Himmler's speech at Poznan: "We had the moral right, we had the duty to our people to destroy this

---

*Cheese maggots (the larvae of the fly *Piophila casei*) are called "skippers" because of their ability to "skip" up to six inches into the air.

people. . . . We do not want, in the end, because we have destroyed a bacillus, to be infected by this bacillus and to die. I will never stand by and watch while even a small rotten spot develops or takes hold." Think of the swarms of rats in *The Eternal Jew*, the characterization of Tutsis as cockroaches, and the image of Armenians as tubercular bacilli.

Because the horror of unclean animals is linked to concerns about cleanliness and purity, and the concepts of cleanliness and purity have a powerful moral resonance, this variety of dehumanization often has a hefty moralistic component, as is exemplified in Himmler's remark that the German people had "the moral right . . . the duty" to exterminate the Jews of Europe. The idea of mass killing as an act of moral cleansing is a common genocidal fantasy. The metaphorical connection between physical and moral filth also explains why this form of dehumanization is often associated with religiously motivated violence. An example can be found in the ferocious clashes between Catholics and Protestants during the sixteenth century, in which each side accused the other of spreading filth and pollution. Natalie Zemon Davis points out in her fine study of religious riots in sixteenth-century France that "The word 'pollution' is often on the lips of the violent, and the concept serves well to sum up the dangers which rioters saw in the dirty and diabolical enemy. . . . For Catholic zealots extermination of the heretical 'vermin' promised the restoration of unity to the body social. . . ."[41]

Thinking of people as vermin isn't limited to genocidal and religious violence. It's also a feature of ordinary war, especially in situations where there is indiscriminate killing. Thus, J. Glenn Gray tells us that in war the enemy is often "considered to be a peculiarly noxious kind of animal toward whom one feels instinctive abhorrence."[42] The language of war contains many such examples. During the Gulf War, U.S. pilot Col. Richard Wright described the attack by U.S. aircraft on Iraqi supply lines using terms reminiscent of the Rwandan genocide. He said, "It's almost like you flipped on the light in the kitchen at night and the cockroaches start scurrying, and we're killing them." And just after the first battle of Fallujah, U.S. General Richard Myers described the Iraqi city as "a huge rat's nest" that was "festering" and therefore needed to be "dealt with"—a discourse that conjured up disturbing images of exterminating filthy vermin.[43]

Imagining dehumanized people as predators presents quite a different picture. Predators have haunted the human imagination since prehistoric times. That we find them both absorbing and terrifying is evidenced by the box office success of films like *Jaws*, *Jurassic Park*, and *War of the Worlds*. There is a powerful biological subtext to our fascination with these monsters. "Great and terrible flesh-eating beasts," writes science journalist David Quammen, "have always shared landscape with humans."

> They were part of the ecological matrix within which *Homo sapiens* evolved. They were part of the psychological context in which our sense of identity as a species arose. They were part of the spiritual systems that we invented for coping. The teeth of big predators, their claws, their ferocity and their hunger, were grim realities that could be eluded but not forgotten. Every once in a while, a monstrous carnivore emerged like doom from a forest or a river to kill someone and feed on the body. . . . Among the earliest forms of human self-awareness was the awareness of being meat.[44]

Our ancestors lived in dread of creatures poised to devour them. The omnipresent possibility of being eaten alive colored their vision of the cosmos, and left an indelible stamp on their cultures. As Barbara Ehrenreich remarks:

> Probably the single most universal theme of mythology is that of the hero's encounter with the monster that is ravaging the land or threatening the very foundations of the universe: Marduk battles the monster Tiamat; Perseus slays the sea monster before it can devour Andromeda; Beowulf takes on the loathsome, night-feeding Grendel. A psychiatrist might say that these beasts are projections of the human psyche, inadmissible hostilities deflected toward mythical

targets. But it might be simpler, and humbler, on our part to take these monsters more literally: as exaggerated forms of a very real Other, the predator beast which would at times eat human flesh.[45]

Because of the danger that they posed to human life, predators are traditionally associated with evil. Medieval Christians conceived of hell as a scene of rampant predation. Here's how one twelfth-century Irish knight described the punishment of the damned.

> Fiery dragons were sitting on some of them and were gnaw-ing them with iron teeth, to their inexpressible anguish. Oth-ers were the victims of fiery serpents, which, coiling round their necks, arms, and bodies, fixed iron fangs into their hearts. Toads, immense, and terrible, also sat on the breasts of some of them, and tried to tear out their hearts with their ugly beaks.[46]

In medieval iconography the gates of hell are portrayed as an animal's gaping maw down the souls of the damned (sometimes specifically as the jaws of a crocodile). Augustine, after all, had proclaimed that "the sinner has been handed over as food for the Devil."[47] This is why when people are dehumanized in accord with the predatory trope they are seen as evil, demonic, bloodthirsty, and even cannibalistic. As we saw in Chapter Three, predatory images like these formed part of the stereo-type of Native Americans as "wild Indians" and "bloodthirsty savages." They were also crucial components of medieval Christians' picture of Jews. In the art and literature of the Middle Ages, "Jews were given horns, tails, a goat's beard (the goat was seen as Satan's disguise), and a noxious odor revealing their descent from the devil. . . ."

> In passion plays Jews were portrayed as evil demons with horns and tails gleefully and sadistically torturing Jesus as he carried the cross and then mutilating his crucified body. In other plays Jews were shown wearing grotesque

costumes, stabbing the Holy Communion, desecrating holy images, conspiring with the devil, and raving like mad dogs. To the medieval mind Jews were not just evil, they were also dangerous and fearful murderers and demons: They slew Christian children to obtain their blood for ancient rituals; armed by Satan with occult powers, they plotted to destroy Christendom and thwart the divine plan.[48]

The response to predators is one of terror, rather than horror. The enemy is ferocious, relentless, formidable, and must be killed in self-defense. For example, in one of the earliest references to the use of dogs in warfare, Polyaenus, the second-century author of *Stratagems in War*, wrote:

When the monstrous and the bestial Cimmerians made an expedition against him, Alyattes brought out for battle his strongest dogs along with the rest of his force. The dogs set upon the barbarians as if they were wild animals, killed many and forced the rest to flee shamefully.[49]

When an anonymous eleventh-century Syrian poet wrote "I know not whether my native land be a grazing ground for wild beasts or yet my home!" with reference to the Christian army that reduced the city of Ma'arra to a heap of smoldering rubble and cannibalized its inhabitants, he voiced a sentiment that was felt by many of his compatriots. His contemporary, the Syrian poet-diplomat Usama Ibn Munqidh, confirmed that the view of the crusaders as animals was quite widespread. "All those who were well-informed about the *Franj* [crusaders]," he wrote, "saw them as beasts superior in courage and in fighting ardour but in nothing else, just as animals are superior in strength and aggression."[50]

If you are in any doubt about whether the conception of the enemy-as-predator is still relevant today, thumb through the pages of Sam Keen's book *Faces of the Enemy*, which shows page after page of propaganda posters representing the enemy as a whole menagerie of carnivorous beasts including tigers, bears, wolves, gigantic spiders, and immense octopuses, or consider the newspaper headlines that I discussed in

Chapter One.[51] Whenever words like *evil, wild,* or *bloodthirsty* start to pepper political discourse, you can be sure that predatory dehumanization is lurking close by.

Finally, we come to the representation of dehumanized people as prey. The image of prey (and ourselves as predators) comes from our ancient legacy of hunting. There is much dispute about exactly when hunting became a feature of human life. In all likelihood, our common ancestor with the chimpanzee hunted, as chimpanzees do today, but the prey that they killed would have provided only a small pant of their diet. However, all parties agree that prehistoric *Homo sapiens* were accomplished hunters, and that hunting played a central role in their lives, as it does in the lives of many tribal people to this day. There is an obvious metaphorical resonance between warfare and hunting—and therefore a tendency to identify warriors with predatory animals and their enemies as prey. This relationship is displayed, for example, in the symbolic architecture of Homer's *Iliad*, which describes how, when Hector entered the fray, "Foam appeared around his mouth and his eyes glowed under his shaggy brows. . . . He came against them like a destructive lion on cattle." Likewise Achilles, the golden boy of the epic, is described as a predatory beast; he is a "raw meat eater" raging "against the flocks of men to make a feast." In contrast, Spartan king Menelaus is depicted as a parasitic fly "beat him off of your skin as often as you like, he goes on biting, and human blood is his dainty dish," while the Greek commanders are like wolves.

> And like the wolves who eat raw flesh, in whose hearts the fury is boundless—who have killed a big-horned stag in the mountains and lap the dark surface of a deep and dark spring with their thin tongues, belching forth blood and gore, with the hearts in their chests dauntless and their bellies glutted—just so did the leaders and rulers of the Myrmidons rush out. . . . [52]

Among the Vikings, would-be elite warriors, or *berserkers*, had to identify with raging bears and go "berserk" (a word that means "dressed in bear-skin") in battle, howling like animals and biting their shields, and Tahitian warriors were exhorted to mimic "the devouring wild dog."[53]

However, dehumanizing the enemy as prey goes well beyond the metaphorical. White settlers in Australia and New Zealand considered the indigenous people as game. Eyewitness Augustus Cutlack described how, during the 1873 Australian gold rush, "Many were the shooting-parties formed and as there was no game to kill it consisted of making repeated attacks on the blacks." Another eyewitness reported in 1889, "There are instances when the young men . . . have employed the Sunday in hunting the Blacks, not only for some definite purpose, but also for the sake of sport" (the turn-of-the-twentieth-century Australian politician King O'Malley argued in parliament that there is no scientific evidence that aborigines are human beings). Similarly, Sir Arthur Gordon, the governor of New Zealand, noted that he had heard "men of culture and refinement" talk of "the individual murder of natives, exactly as they would talk of a day's sport, or of having to kill some troublesome animal."[54]

Just as hunters preserve trophies of the animals they have killed, warriors have been known to take souvenirs from their human quarry. According to the Greek historian Herodotus, Scythian warriors took the heads of the men they killed in battle and then scalped them. "The Scyth is proud of these scalps," he remarks, "and hangs them from his bridal-rein; the greater the number of such napkins a man can show, the more highly he is esteemed among them."

> Many make themselves cloaks . . . by sewing a quantity of these scalps together. Others flay the right arms of their dead enemies, and make of the skin, which is stripped off with the nails hanging to it, a covering for their quivers.[55]

It was once thought that the practice of scalping was introduced to the New World by European colonists. But archaeological evidence shows that Native Americans took scalps and other trophies long before Columbus landed in the Caribbean.

The removal of heads, scalps, eyes, ears, teeth, cheekbones, mandibles, arms, fingers, legs, feet, and sometimes genitalia for use as trophies by Amerindians was an ancient and widespread practice in the New World. Some groups in Colombia and in the Andes kept the entire skins of dead enemies.[56]

In most cases, we don't know whether dehumanization played a role in these grizzly practices, but we do have information about the role of dehumanization in the headhunting practices of some cultures, one of which is the Mundurucú of Brazil. The Mundurucú were once an extremely warlike tribe, who regarded all outsiders as enemies. Robert F. Murphy, an anthropologist who lived among them, has noted that "war was considered an essential and unquestionable part of their way of life, and foreign tribes were attacked because they were enemies by definition." Attacks on other villages took the form of headhunting raids, and in these raids "the enemy was looked upon as game to be hunted, and the Mundurucú still speak of the *pariwat* (non-Mundurucú) in the same terms that they reserve for peccary and tapir."[57] Closer to home, it's well known that during the Vietnam War, U.S. servicemen sometimes removed the ears from dead Vietnamese and kept them as trophies (sometimes stringing them to make necklaces). Vietnam veteran Jon Neely recounts how he approached a corpse and "reached down and cut one of the guy's ears off and poked a hole in it and hung it on a chain."

> I had become, I don't know, part animal I guess you could call it. . . . I was enjoying the firefights and enjoying the killing, and at one time I displayed as many as thirteen ears on this chain that I had hanging off my gear. I look back on it now and I wonder to myself, Jeeze, what the heck happened to me?[58]

I mentioned in Chapter One that similar atrocities were committed in the Pacific theater during World War II, and that they were probably related to Americans' dehumanization of the Japanese, who were held to be "really subhuman, little yellow beasts. . . ."[59] This happened often enough for the commander in chief of the Pacific fleet to issue an order that "No part of the enemy's body may be used as a souvenir."[60] Paul Fussell recalls visiting an ex-marine who had fought on Guadalcanal. Upon hearing that Fussell was working on a book about World War II, the veteran produced a shoebox full of snapshots, some of which were pictures of Japanese trophy skulls. One showed a skull mounted on a pole, another showed one displayed on a ruined Japanese tank, and the third was "being boiled in a metal vat, and two marines were busy poking it and turning it with sticks." One of the two marines was Fussell's host.

> My friend assured me that securing and preserving Japanese skulls was by no means a rare practice, and that it had started on Guadalcanal, at the virtual beginning of the ghastly fighting in the Pacific. Because the marines had not yet learned the full depths of Japanese ruthlessness, this early skull-taking seems to register less a sinking of the U.S. Marine Corps to the Japanese level of brutality—that would come later—than a simple 1940s American racial contempt. Why have more respect for the skull of a Jap than for the skull of a weasel, a rat, or any other form of mad, soulless vermin?[61]

There's something especially disturbing about people being hunted as game. The aversion to it stems from our awareness of the fact that this is killing for the sheer pleasure of it, without any element of self-defense. This is the sort of violent engagement that evokes the "combat high" described earlier in this chapter and explains the attraction of war porn. James Hebron, a U.S. Marine scout-sniper during the Vietnam War, expresses this atavistic impulse with chilling frankness.

> That sense of power, of looking down the barrel of a rifle at somebody and saying, "Wow, I can drill this guy." Doing it is something else too. You don't necessarily feel bad; you

feel proud, especially if it's one on one. It's the throw of the hat. It's the thrill of the hunt.[62]

We're now close to the end of the journey. In the next, concluding chapter, I'm going to summarize the major points that I have made throughout the book, and then address one last question.

9

QUESTIONS FOR A THEORY OF DEHUMANIZATION

Having reached the end of his journey, the author must ask his readers' forgiveness for not being a more skillful guide and for not having spared them empty stretches of road and troublesome detours. There is no doubt that it could have been done better.

—SIGMUND FREUD, *CIVILIZATION AND ITS DISCONTENTS*[1]

I'VE TRIED TO COVER A LOT of territory in this book, and I am left wondering how successful I've been at getting the most important points across. So, to bring the story to a conclusion, I'm going to revisit the questions that I raised in Chapter One—questions that I said any adequate theory of dehumanization should address. I've covered all but one of them in the intervening chapters, so what I say about them here will amount to a précis. However, there's one question that I haven't discussed at all. So, in the final part of this chapter, I'm going to offer some reflections on what is arguably the weightiest issue of them all— the question of what we can do about dehumanization.

What does it mean to think of someone as a human being, and what is it, exactly, that dehumanized people are supposed to lack? Thinking of someone as a human being is thinking of that person as a being with a human essence: an imaginary "something" that all humans are supposed to possess, and which makes them human. A dehumanized person is thought to lack this essence. They are thought of as humanoid or quasi-human beings—as human in appearance only.

263

What sorts of creatures are dehumanized people imagined to be? Dehumanized people are imagined as subhuman animals, because they are conceived as having a subhuman essence. Even though such people have a human form, this is deceptive, because "inside" they are really something else. They are imagined to have the essence of creatures that elicit negative responses, such as disgust, fear, hatred, and contempt, and are usually thought of as predators, unclean animals, or prey.

What is it about the human mind that enables us to conceive of people as less than human? Our ability to conceive of others as subhuman depends on five features of our psychology. We must have a domain-specific cognitive module for folk-biology, because this is what causes us to intuitively divide up the biological world into natural kinds that we call "species." Second, we've got to have a domain-specific cognitive module for folk-sociology that carves up the human world into the natural kinds called "races." Third, we've got to be capable of engaging in second-order thought, which enables us to reflect on our concepts of species and races. Fourth, we have to conceive of biological species and human races as having unique essences that make them what they are—essences that are distinct from how they appear, that are transmitted from parents to offspring, and so on. And fifth, we need to embrace the idea of a "thick" hierarchy of natural kinds—some version, however crude, of the idea of the great chain of being.

How does dehumanization work, why does it occur, and what function does it serve? Dehumanization is a response to conflicting motives. It occurs in situations where we want to harm a group of people, but are restrained by inhibitions against harming them. Dehumanization is a way of subverting those inhibitions. For a population to be dehumanized they have to be perceived as a race (a natural human kind) with a unique racial essence. The racial essence is then equated with a subhuman essence, leading to the belief that they are subhuman animals. The function of dehumanization is to override inhibitions against committing acts of violence.

Is the dehumanizing impulse universal, or is it culturally and historically specific? Is dehumanization a hard-wired product of our biological evolution, or is it acquired? Nobody knows if dehumanization is universal, but it is very widespread. Although it has a form that cuts across cultures, its content in any given case is culturally determined. Dehumanization is *not* a biological adaptation. It wasn't put in place by natural selection and it's not hard-wired. It's an unconscious strategy for dealing with psychological conflict.

Now we come to the question that I have not addressed. *What can be done about the problem of dehumanization?* Given the historical role of dehumanization as the handmaiden to war, genocide, and slavery, it's obvious that preventing it would be incalculably beneficial. The main problem confronting anyone wishing to address this question is that dehumanization has barely been studied, and consequently very little is known about it. So, we don't yet have a knowledge base from which to derive practical policies and interventions for limiting or eliminating dehumanization. Given that we don't now have answers to this pressing question, we need to take one step back and think about what sort of approach is likely to yield such answers. In this book I have assumed that a broadly scientific approach to the phenomenon of dehumanization is the only sensible one. However, there are people who believe that scientific knowledge is irrelevant to developing strategies for combating dehumanization (there are even those who believe that a scientific approach *fosters* dehumanization by turning human beings into objects of empirical scrutiny). Many such people hold that we know enough on the basis of common sense and that science has nothing to add to this picture, even in principle. In the remainder of this chapter, I'm going to discuss two nonscientific stances toward dealing with the problem of dehumanization, and will argue that neither of them is satisfactory.

My goal is to defend a broadly scientific approach to the problem of dehumanization because this is the only approach that has any chance of succeeding. I am not claiming that science can provide a "cure" for dehumanization, or that a scientific understanding of the dehumanizing process is bound to produce uniformly beneficial results. In fact, I think that any such knowledge is potentially hazardous. That this is more than a bare possibility is suggested by an anecdote in Peter Watson's book *War on the Mind*. Watson describes a U.S. Navy project for preparing elite commando units to cope with the "stress of killing" in which an important part of the regimen was training recruits to dehumanize the enemy.

> In this last phase [of the training] the idea is to get the men to think of the potential enemies that they will have to face as inferior forms of life. They are given lectures and films which portray personalities and customs in foreign countries whose interests may go against the USA. The films are biased to present the enemy as less than human: the stupidity of local customs is ridiculed, local personalities are presented as evil demigods rather than legitimate political figures.[2]

This is heavy-handed stuff, but it's not difficult to imagine that a more sophisticated understanding of how dehumanization works could be exploited for indoctrinating more effective killers. Knowledge is a double-edged sword. If science can yield insights into methods for combating dehumanization, it may also suggest strategies for cultivating dehumanization more effectively. Knowledge is powerless to make people less destructive. However, it can provide the tools to help them become less destructive, if that is what they desire.

Richard Rorty, the late distinguished American philosopher, was one of the very few thinkers to have directly addressed the problem of dehumanization. His essay, "Human Rights, Rationality, and Sentimentality," originally delivered as an Amnesty International lecture in 1993, provides a useful springboard for thinking about the options.[3]

The essay kicks off with an excerpt from an article by David Rieff published in *The New Yorker* in November 1992 about the horrific events that were unfolding in Bosnia. "To the Serbs, the Muslims are no longer human," Rieff observes, ". . . Muslim prisoners, lying on the ground in rows, awaiting interrogation, were driven over by a Serb guard in a small delivery van."

> A Muslim man in Bosanski Petrovac . . . [was] forced to bite off the penis of a fellow-Muslim. . . . If you say that a man is

not human, but the man looks like you and the only way to identify this devil is to make him drop his trousers—Muslim men are circumcised and Serb men are not—it is probably only a short step, psychologically, to cutting off his prick. . . . There has never been a campaign of ethnic cleansing from which sexual sadism has gone missing.[4]

The moral that Rorty extracts from this harrowing account is that "Serbian murderers and rapists do not think of themselves as violating human rights. For they are not doing these things to fellow human beings, but to *Muslims*."

> They are not being inhuman but rather are discriminating between true humans and pseudohumans. They are making the same sort of distinction that the Crusaders made between humans and infidel dogs, and Black Muslims make between humans and blue-eyed devils. The founder of my university was able both to own slaves and to think it self-evident that all men were endowed by their creator with certain inalienable rights. He had convinced himself that the consciousness of Blacks, like that of animals, "participate[s] more of sensation than reflection." Like the Serbs, Mr. Jefferson did not think of himself as violating *human* rights.[5]

He further observes that we are inclined to look upon the purveyors of violence as demonic, as monsters, or as vicious animals. So, although it stops short of atrocity, our attitude toward *them* is alarmingly like the attitude that *they* take toward their victims. But there is worse to come. Rorty argues that we are as prone to dehumanize the victims of brutality as we are the perpetrators. "We think of Serbs or Nazis as animals," he observes, "because ravenous beasts of prey are animals. We think of Muslims or Jews being herded into concentration camps as animals, because cattle are animals. Neither sort of animal is very much like us, and there seems no point in human beings getting involved in quarrels between animals."[6] This, I think, is an important insight that may explain our tolerance for so-called collateral damage in foreign military interventions and our callousness in the face of human rights violations on foreign soil (or those perpetrated against immigrants, minorities, and the poor on our own soil). I am reminded of the widespread obliviousness to the seemingly endless civil war in the Democratic Republic of the Congo, and the horrendous plague of rape that has accompanied it. In an article entitled "No, sexual violence is not 'cultural,'" published in *The New York Times* on June 25, 2010, Lisa Shannon points out that many people in the developed world falsely and self-servingly assume that the epidemic of rape is a "traditional" feature of Congolese culture. "Describing the violence in Congo as 'cultural,'" she notes, "is more than offensive. It is dangerous."

Appealing to the voice of reason is a time-honored response to such brutalities. This approach has been implicitly or explicitly advocated by most moral philosophers from Plato, through Kant, right up to their present-day heirs. According to this rationalistic view, dehumanization is a symptom of ignorance, and is to be cured by administering an appropriate dose of intellectual enlightenment. One is to convince the dehumanizer that there's something about being human that makes all of us worthy of a kind of respect that's incompatible with perpetrating atrocity. Human nature contains a special ingredient—variously described as rationality, sentience, a soul, and so on—that is absent in other animals, and it is this special ingredient that underwrites human rights. So, those who believe that doing violence to others is licensed by their race, religion, or nationality are simply failing to recognize a deep truth about what it is to be human.

Rorty points out that this approach is both misguided and ineffectual because it begs the question of who should be counted as human. The merchants of horror discussed by Rieff might endorse the idea of human rights, *while denying that Muslims are human.* Suppose that you were confronted with a Serbian who believes that Muslims are less than human, and that you wanted to set him right. What line of reasoning could you use to convince him that Muslims are human? You might point out that Serbs and Muslims are members of the same species, and that Serbs are human, so Muslims must be human, too. But this relies on the blatantly false premise that two individuals who are members of the same species must be identical in every respect. Your interlocutor could point out that some people have blue eyes and others don't. So, why can't it be that some people are human and others are not? "Ah," you might respond, "but this principle doesn't apply to superficial characteristics like eye color. It's about our *essential* nature." But in saying this you've exposed the poverty of your argument, because the idea that all *Homo sapiens* are essentially human is precisely what's at issue.

As I mentioned in Chapter Three, science can't be recruited to shore up the rationalist's argument. Biologists tell us that we are all members of the same species, *Homo sapiens,* but it's beyond their remit to say that all *Homo sapiens* are human. *Human* isn't a scientific concept at all. It's a folk-concept that means, roughly, *one of us.* As Rorty insightfully observes, such people "are *morally* offended . . . by the suggestion that they treat people whom they do not think of as human as if they were human."

> When utilitarians tell them that all pleasures and pains felt by members of our biological species are equally relevant to moral deliberation, or when Kantians tell them that the ability to engage in such deliberation is sufficient for membership in the moral community, they are incredulous. They rejoin that these philosophers seem oblivious to blatantly obvious moral distinctions, distinctions any decent person would draw.7

Rorty thinks that advocates of the rationalistic approach have been barking up the wrong tree. Rather than looking for explanations for why all people deserve to be treated with compassion and respect, we ought to be working at *creating a world* in which people are treated with compassion and respect. Human rights aren't lying around waiting to be discovered. They're made, not found. But how can this be accomplished? He suggests that we should take our cue from Hume. Morality is about feeling, so if we want people to treat one another humanely we ought to be appealing to their feelings instead of offering them dry theoretical arguments. We need to help people get to know one another by telling them "long, sad, sentimental stories." "Such stories," Rorty observes, "repeated and varied over the centuries, have induced us, the rich, safe, powerful people, to tolerate and even to cherish powerless people—people whose appearance or habits or beliefs at first seemed an insult to our own moral identity, our sense of the limits of permissible human variation." These stories, he thinks, will make people "less tempted to think of those different from themselves as only quasi-human." He explains, "The goal of this sort of manipulation of sentiment is to expand the reference of the terms 'our kind of people' and 'people like us.'" In other words, its aim is to expand the reference of the term *human* to include *everyone*.[8] I call this the *sentimentalist* approach.

The sentimentalist strategy has a greater chance of being effective than the rationalistic one does. Hume was right; we're moved by passion rather than reason, so it seems to follow that if we want to move others, we need to appeal to their passions to steer them in the right direction. The *right* direction? That word "right" points to a problem— actually, a danger—that inheres to the sentimentalist project. Dehumanizing stories are among the most powerful and moving ones. And they are often long, sad, sentimental stories that evoke floods of sympathy for those who suffer at the hands of *animals in human form*. Had Rorty looked behind the headlines he would have discovered that the brutalities that Rieff described were inspired by years of state-sponsored propaganda, propaganda that told stories about the sufferings of the Serbian people at the hands of their wicked Muslim enemy. The official Serbian narrative was that Muslims were perpetrating genocide on innocent Serbians, and all those who opposed Milosevic's regime were complicit in that genocide. That's why Milosevic responded to trade sanctions against Serbia by telling a long, sad, sentimental story:

> I do not know how you will explain to your children, on the day when they discover the truth, why you killed our children, why you led a war against three million of our children, and with what right you turned twelve million inhabitants of

Europe into a test site for the application of what is, I hope, the last genocide of this century.[9]

Hitler told a sentimental story about the sufferings of the Aryan race at the hands of the Jewish vermin; Rwandan Hutus told a sentimental story about their sufferings at the hands of the Tutsi cockroaches; and American white supremacists told a sentimental story (adapted for cinema as *The Birth of a Nation*) about the sufferings of white Southerners at the hands of their bestial former slaves. Furthermore, Rorty's explanation of why people dehumanize one another is both oddly simplistic and transparently false. He writes that people who dehumanize others do so because they are deprived of "security and sympathy."

> By "security" I mean conditions of life sufficiently risk-free to make one's differences from others inessential to one's self-respect, one's sense of worth. These conditions have been enjoyed by North Americans and Europeans—the people who dreamed up the human rights culture—much more than they have been enjoyed by anyone else. . . . Security and sympathy go together, for the same reasons that peace and economic productivity go together. The tougher things are, the more you have to be afraid of, the more dangerous your situation, the less you can afford the time or effort to think about what things might be like for people with whom you do not immediately identify.[10]

Do Americans and Europeans *really* corner the market on sympathy? Tell that to the survivors of Auschwitz. Tell it to the descendants of American slaves and to Native Americans. Tell it to the Herero of Namibia and the millions butchered under King Leopold's regime in the Congo Free State.

Why this retreat into trite, self-congratulatory ethnocentrism? I think that it's best explained by Rorty's refusal to countenance the existence of human nature. "There is a growing willingness," he remarks, with evident approval, "to neglect the question 'What is our nature?' and to substitute the question 'What can we make of ourselves?'"

> We are much less inclined than our ancestors were to take ontology or history or ethology as a guide to life. We are much less inclined to pose the ontological question "What *are* we?" because we have come to see that the main lesson of both history and anthropology is our extraordinary malleability. We are coming to think of ourselves as the flexible, protean, self-shaping animal rather than as the rational animal or the cruel animal.[11]

If you neglect the question "What is our nature?" then you can't look for sources of dehumanization in our nature, and you have no alternative but to gravitate toward a shallow social determinism. I don't deny that social conditions are vital for explaining particular instances of dehumanization. I've made the point repeatedly in the pages of this book. But social construction presents only a slice—and sometimes a very thin slice—of the truth. In opposing the question "What are we?" to the question "What can we make of ourselves?" Rorty offers up a false dichotomy. What we can make of ourselves is constrained by what we are for the same reason that what a sculptor can make out of a block of stone is constrained by the properties of the stone. To work the stone effectively the sculptor must understand its properties. He must know what to do at what point and with what tools. By the same token, we self-sculptors have to understand the properties of human nature if we are to have any hope of getting the results that we're aiming at. Even if you believe, with Rorty, that creating a better world depends on telling better stories, the fact that certain kinds of stories reliably produce certain kinds of effects requires an explanation, and any adequate explanation must tell us why human animals are disposed to respond to *that* kind of story in *that* kind of way.

To deal effectively with dehumanization, we need to understand its mechanics. There's simply no viable alternative. To do this, we need to bring science to bear on those aspects of human nature that sustain the dehumanizing impulse. I've made a few suggestions in this book, but my efforts are only a start. The study of dehumanization needs to be made a priority. Universities, governments, and nongovernmental organizations need to put money, time, and talent into figuring out exactly how dehumanization works and what can be done to prevent it. Maybe then we can use this knowledge to build a future that is less hideous than our past: a future with no Rwandas, no Hiroshimas, and no Final Solutions.

Can this be done? Nobody knows, because nobody's ever tried.

APPENDIX I

PSYCHOLOGICAL ESSENTIALISM

1. Beliefs about essences are intuitive and need not be explicit.
2. Essences are imagined to be shared by members of natural kinds—kinds that are discovered rather than invented, real rather than merely imagined, and rooted in nature.
3. The contents of essentialist beliefs are sensitive to cultural norms.
4. Essences give rise to the stereotypical features associated with a kind. Deviations from the stereotype indicate that something is preventing the essence from being expressed or distorting its expression.
5. Essences are inherent and unalterable. An item can't lose or change its essence while retaining its identity.
6. Essences are absolute rather than incremental—there are no degrees of having an essence.
7. Essences are transferred from parent to offspring or from host to client.
8. Essences are not conserved—the transfer of an essence does not diminish the quantity of the essence remaining in the parent or donor.
9. Essences remain stable across transformations. Changes in appearance do not correspond to changes in essence.

APPENDIX II

PAUL ROSCOE'S THEORY OF DEHUMANIZATION IN WAR

My book *The Most Dangerous Animal: Human Nature and the Origins of War* was published in August 2007. In its penultimate chapter, I proposed a crude version of the theory set out in the present book. A month later, University of Maine anthropologist Paul Roscoe published a paper in *American Anthropologist* advancing a theory that was amazingly similar to mine. Roscoe and I had been working on the same problem, and had independently reached very similar conclusions.[1]

Roscoe begins with Richard Wrangham's hypothesis that men have a disposition to seek out low-cost opportunities for killing outsiders. He then goes on to argue that we also have a strong aversion to taking human life, illustrating the point with examples from Christopher Browning's book *Ordinary Men: Reserve Police Battalion 101 and the Final Solution in Poland*.[2] Browning tells the story of how ordinary, middle-aged German men were drafted into Police Battalion 101 to perform mass executions of Jews in Poland. These "ordinary men" massacred at least 38,000 men, women, and children, and were involved in the deportation of another 45,000 to death camps. Roscoe comments, quoting Browning, that:

> If these men were motivated to seek out low-cost opportunities
> to kill, in sum, we should expect them to have participated

eagerly in these massacres. Most of them appear to have experienced a marked aversion, at least to begin with. . . . But
between 10 and 20 percent of the unit avoided killing by re-
questing that they be excused from execution details, by
sidling to the back when execution squads were mustered, or
by spreading the word that they were "too weak" for such
work. . . . Of the remainder, most "did not seek opportunities to kill (and in some cases, refrained from killing, contrary to standing orders, when no one was monitoring their
actions)". . . . Of special note, "almost all of them—at least
initially—were horrified and disgusted by what they were
doing."[3]

How, then, can we account for our propensity to kill our own kind?
Roscoe argues that the evolution of intelligence made us capable of
subverting our inhibitions. "Under this hypothesis," he writes, "the
stage was set for humans to become a killer species when they or their
predecessors became sufficiently intelligent to recognize when it was
advantageous to kill."[4] As our ancestors became more intelligent they
devised strategies to "short-circuit" their inhibitions against killing.
One method was to invent long-range weapons that insulate the killer
from his own actions. Another is to ingest mind-altering drugs to distort
realistic perception. Other strategies include shifting responsibility onto
supernatural beings who supposedly command one to go to war, denigrating a group of people to legitimize acts of violence against them, and
using drumming or chanting to induce altered states of consciousness.
However:

The most common way to overwhelm an aversion to killing . . . is to combine dehumanization of the enemy, which
denies him or her conspecifics status, with an image that
elicits killing responses appropriate toward nonhuman species.
    Frequently, war is depicted as hunting rather than
murder, and the enemy as a game animal rather than a

human. . . . Alternatively enemies are depicted as enraged or unreasoning micro- or macropredators—bacilli, parasites, disease-spreading vermin, snakes, large carnivores, or capricious demons—agents that represent an imminent threat to survival and so incite a lethal reaction. . . . [5]

Paul Roscoe and I see eye to eye on almost every major point. Our main disagreement is about timing. Drawing on Jane Goodall's observation (mentioned in Chapter Two) that chimpanzees seem to "dechimpize" other chimps when attacking them, he argues that the capacity for dehumanization may date from before the split between the chimpanzee and human lineages, whereas I believe that it emerged much later.

NOTES

PRELUDE: CREATURES OF A KIND SOMEWHAT INFERIOR

1. For a penetrating discussion on the extension of "the human" in eighteenth-century thought, see C. W. Mills, "Kant's *Untermenschen*," in *Race and Racism in Modern Philosophy*, ed. Andrew Valls (Ithaca, NY: Cornell University Press, 2005).
2. Mills, "Kant's *Untermenschen*," 3.
3. J. Philmore, *Two Dialogues on the Man-Trade* (London: Waugh, 1760), 12.
4. I know of only four books on the topic of dehumanization. They are Sam Keen's *Faces of the Enemy: Reflections of the Hostile Imagination* (Minnetonka, MN: Olympic Marketing Corporation, 1986), William Brennan's *Dehumanizing the Vulnerable: When Word Games Take Lives* (Chicago: Loyola University Press, 1995), Linda LeMoncheck's *Dehumanizing Women: Persons as Sex Objects* (Totowa, NJ: Rowman and Allanheld, 1985), and Leonard Cassuto's *The Inhuman Race: The Racial Grotesque in American Literature and Culture* (New York: Columbia University Press, 1997). Keen's book was a groundbreaking study of images of the enemy in propaganda. It includes a number of striking visual examples of dehumanization, but because of its restrictive purview, and its lack of engagement with the wider scientific and philosophical literature, it cannot be considered a scholarly analysis of dehumanization. Brennan is a Roman Catholic professor of social work, whose book analyzes the dehumanizing language legitimizing violence against seven populations (including human embryos—Brennan is an opponent of abortion), but the book contains very little analysis. LeMoncheck's book is a substantial contribution to the feminist literature, but she means something quite different by "dehumanization" than I do. Cassuto's book is a fine study of dehumanization and racism in American culture from the vantage point of a literary scholar.
5. I. Eibl-Eibesfeldt, *Human Ethology* (New York: Aldine de Gruyter, 1989), 403.

6. LeMoncheck, *Dehumanizing Women*: A. Cahill, *Rethinking Rape* (Ithaca NY: Cornell University Press, 2001); A. Dworkin, "Against the Male Flood: Censorship, Pornography, and Equality," in *Oxford Readings in Feminism: Feminism and Pornography*, ed. D. Cornell (New York: Oxford University Press, 2000); C. MacKinnon, *Feminism Unmodified* (Cambridge, MA: Harvard University Press, 1987). For critical responses, see M. Nussbaum, "Objectification," *Philosophy and Public Affairs* 24, no. 4 (1995): 249–91 and A. Soble, *Pornography, Sex, and Feminism* (New York: Prometheus, 2002).

7. R. M. Brown, "1492: Another Legacy: Bartolomé de Las Casas—God Over Gold in the Indies," *Christianity and Crisis* 51 (1992): 415.

1. LESS THAN HUMAN

1. The nursery rhyme is from N. Nazzal & L. Nazzal, "The Politicization of Palestinian Children: An Analysis of Nursery Rhymes," in *Islamophobia and Anti-Semitism*, ed. H. Schenker and Z. Abu-Zayad (Jerusalem: Palestine-Israel Journal, 2006), 161. The comment by Rabbi Ovadia Yosef, which originally appeared in *Haaretz* (March 20, 2000), is from Sivan Hirsch-Hoefler and Eran Halperin, "Through the Squalls of Hate: Arabic-phobic Attitudes Among Extreme Right and Moderate Right in Israel," in Schenker and Abu-Zayad, *Islamophobia and Anti-Semitism*, 103.

2. C. Hedges, *War Is a Force That Gives Us Meaning* (New York: Anchor, 2003), 94.

3. Ibid., 95.

4. C. McGreal, "Hamas Celebrates Victory of the Bomb as Power of Negotiation Falters," *The Guardian*, September 12, 2005.

5. Steven Erlanger, "In Gaza, Hamas Insults to Jews Complicate Peace," *New York Times*, April 1, 2008. The Al-Jaubari excerpt is from "To Disclose the Fraudulence of the Jewish Men of Learning," in *The Legacy of Islamic Anti-Semitism: From Sacred Texts to Solemn History*, ed. A. B. Bostom (New York: Prometheus, 2008), 321.

6. G. J. Annas and M. A. Grodin (eds.), *The Nazi Doctors and the Nuremberg Code: Human Rights in Human Experimentation* (Oxford: Oxford University Press, 1995), 67.

7. Quoted in J. Herf, *The Jewish Enemy: Nazi Propaganda During World War II and the Holocaust* (Cambridge, MA: Harvard University Press, 2006), 101. M. Domarus (ed.), *Hitler: Reden und Proklamationen, 1932–1945*, vol. 2; *Untergang, 1939–1945* (Neustadt: Schmidt, 1963), 1967; A. Margalit and G. Motzkin, "The Uniqueness of the Holocaust," *Philosophy and Public Affairs* 25, no. 1 (1996): 65–83.

8. M. R. Habeck, "The Modern and the Primitive: Barbarity and Warfare on the

Eastern Front," in *The Barbarization of Warfare*, ed. G. Kassimeris (New York: New York University Press, 2006), 95.

9. P. Knightly, *The First Casualty: The War Correspondent as Hero and Myth-Maker from the Crimea to Kosovo* (Baltimore: Johns Hopkins, 2002), 188, 269.

10. I. Ehrenberg, "Kill," quoted in A. Goldberg, *Ilya Ehrenburg: Revolutionary, Novelist, Poet, War Correspondent, Propagandist. The Extraordinary Life of a Russian Survivor* (London: Weidenfeld and Nicolson, 1984), 197.

11. G. McDonough, *After the Reich: The Brutal History of the Allied Occupation* (New York: Basic Books, 2007), 26, 46, 50.

12. J. W. Dower, *War Without Mercy: Race and Power in the Pacific War* (New York: Pantheon, 1986), 241.

13. H. Katsuichi, *The Nanjing Massacre: A Japanese Journalist Confronts Japan's National Shame* (Armonk, NY: M. E. Sharpe, 1999), 119–121.

14. D. C. Rees, *Horror in the East: Japan and the Atrocities of World War II* (New York: DaCapo Press, 2002), 28.

15. K. Blackburn, *Did Singapore Have to Fall? Churchill and the Impregnable Fortress* (New York: RoutledgeCurzon, 2003), 94. A. Schmidt, *Ianfu: The Comfort Women of the Japanese Imperial Army of the Pacific War: Broken Silence* (Lewiston, UK: Mellen Press, 2000), 87.

16. The remark about Nazis is cited in M. C. C. Adams, *The Best War Ever: America in World War Two* (Baltimore: Johns Hopkins University Press, 1994), 98. R. Holmes, "Enemy, attitudes to," in *The Oxford Companion to Military History*, ed. R. Holmes (New York: Oxford University Press, 2001), 284.

17. Dower, *War Without Mercy*, 1986. H. Wouk, *The Caine Mutiny: A Novel* (Boston: Back Bay Books, 1992), 27. E. Pyle, *The Last Chapter* (New York: Henry Holt & Co., 1945), 5. Blamey is quoted in Dower, *War Without Mercy*, 71. E. Thomas, *Sea of Thunder: Four Commanders and the Last Naval Campaign, 1941–45* (New York: Simon and Schuster, 2007).

18. C. A. Lindbergh, *The Wartime Journals of Charles A. Lindbergh* (New York: Harcourt Brace Jovanovich, 1970). Dower, *War Without Mercy*, 66.

19. J. G. Gray, *The Warriors: Reflections on Men in Battle* (Lincoln: University of Nebraska Press, 1998), 150.

20. *Leatherneck* 28, no. 3 (March, 1945). Dower, *War Without Mercy*, 91, 40–41, 71, 77–78.

21. S. M. Hersh, "Torture at Abu Ghraib," *New York Times*, May 10, 2004.

22. V. E. Bonnell, *Iconography of Power: Soviet Political Posters Under Lenin and Stalin* (Berkeley, CA: University of California Press, 1998), 221.

23. Unpublished speech, quoted in R. J. Lifton and N. Humphrey, *In a Dark Time* (Cambridge, MA: Harvard, 1984), 10.

24. Knightly, *The First Casualty*. S. Keen, *Faces of the Enemy: Reflections on the Hostile Imagination* (New York: HarperCollins, 1986).

25. Quoted in H. Dabashi, *Post-Orientalism: Knowledge and Power in Times of Terror* (New Brunswick, NJ: Transaction, 2009), x.

26. Quoted in S. Kinzer and J. Rutenberg, "The Struggle for Iraq: American Voices," *New York Times*, May 13, 2004. Quoted in D. Berreby, *Us and Them: Understanding Your Tribal Mind* (New York: Little Brown and Company, 2005), 239.

27. Quoted in BBC News, "Iraq Abuse 'Ordered from the Top,'" June 15, 2004. http://www.news.bbc.co.uk/1/hi/world/americas/3806713.stm and quoted in D. Berreby, *Us and Them*. Keen, *Faces of the Enemy*.

28. "Boortz: "Islam is a deadly virus" and "We're going to wait far too long to develop a vaccine to find a way to fight this," *Media Matters for America*, October 18, 2006 <http://mediamatters.org/items/200610018005>. "Savage nation: It's not just Rush; talk radio host Michael Savage: 'I commend' prisoner abuse 'we need more,'" May 14, 2004 <http://mediamatters.org/items/200160724007>. The above are mentioned in E. Steuter and D. Wills, *At War with Metaphor: Media, Propaganda and Racism in the War on Terror* (New York: Lexington, 2008), 131–155.

29. M. Dowd, "Empire of Novices," *New York Times*, December 3, 2003.

30. E. Steuter and D. Wills, "Discourses of Dehumanization: Enemy Construction and Canadian Media Complicity in the Framing of the War on Terror," presented at the Canadian Communication Association Annual Meeting, Ottawa, Canada, 2009. http://www.mta.ca/faculty/socsci/sociology/steuter/discourses_of _dehumanization.pdf

31. Steuter and Wills, *At War with Metaphor*. Steuter and Wills, "Discourses of Dehumanization," 2009, 69–99.

2. STEPS TOWARD A THEORY OF DEHUMANIZATION

1. According to the *Oxford English Dictionary*, the first known use of the word was in 1818. A. Opsahl, "Technology Shouldn't Dehumanize Customer Service," *Government Technology News*, January 1, 2009. D. Holbrook, *Sex and Dehumanization* (London: Pitman, 1972). N. Dawidoff, "Triathalons Dehumanize," *Sports Illustrated*, October 16, 1989, 71(16). J. P. Driscoll, "Dehumanize at Your Own Risk," *Educational Technology* no. 18: 34–36. S. Hirsh, "Torture of Abu Ghraib," *The New Yorker*, May 10, 2004. For many other examples, see A. Montague and F. Matson, *The Dehumanization of Man* (New York: McGraw-Hill, 1983) and N. Haslam, "Dehumanization: An Integrative Review," *Personality and Social Psychology Review* 10, no. 3 (2006): 252–264.

2. L. LeMoncheck, *Dehumanizing Women: Treating Persons as Sex Objects* (Lanham, MD: Rowman and Littlefield, 1985).

3. A. Dworkin, "Against the Male Flood: Censorship, Pornography, and Equality," in *Oxford Readings in Feminism: Feminism and Pornography*, ed. D. Cornell (New York: Oxford University Press, 2000), 30–31. C. MacKinnon, *Feminism Unmodified* (Cambridge, MA: Harvard University Press, 1987). For a subtle, critical analysis of the concept of objectification, see M. Nussbaum, "Objectification," *Philosophy and Public Affairs* 24, no. 4 (1995): 249–291.

4. H. Holtzer, *The Lincoln-Douglas Debates: The First Complete Unexpurgated Text* (New York: Fordham University Press, 1994), 151.

5. Ibid., 348.

6. C. P. Cavafy, "Waiting for the Barbarians," in *Collected Poems* (Princeton: Princeton University Press, 1992), 18.

7. B. Isaac, *The Invention of Racism in Classical Antiquity* (Princeton: Princeton University Press, 2004), 506.

8. H. Lloyd-Jones, *Females of the Species: Semonides on Women* (Park Ridge, NJ: Noyes, 1975), 36.

9. Isaac, *The Invention of Racism in Classical Antiquity*, 197.

10. Aesop, Fable 515, *Aesop's Fables*, trans. L. Gibbs (New York: Oxford University Press, 2002), 238.

11. For a clear exposition of Aristotle's thinking on this issue, see C. Shields, *Aristotle* (New York: Routledge, 2007). For a more successful, post-Darwinian philosophical account of function, see R. G. Millikan, *White Queen Psychology and Other Essays for Alice* (Cambridge, MA: MIT Press, 1993).

12. Strictly speaking, it's more accurate to say that pure water consists of oxides of hydrogen in the relation 1:2. See I. Hacking, "Putnam's Theory of Natural Kinds and Their Names Is Not the Same as Kripke's," *Principia* 1 (2007): 1–24.

13. H. Putnam, "The Meaning of 'Meaning,'" *Philosophical Papers, Vol. 2: Mind, Language and Reality* (Cambridge, MA: Cambridge University Press, 1975).

14. Although still young, the study of essentialistic thinking is a burgeoning nexus for interdisciplinary research, with a substantial literature. See, for example, W. K. Ahn, et al., "Why Essences Are Essential in the Psychology of Concepts," *Cognition* 82 (2001): 59–69; S. Atran, et al. "Generic Species and Basic Levels: Essence and Appearance in Biology," *Journal of Ethnobiology*, 17 (1997): 22–45; J. Bailinson, et al., "A Bird's Eye View: Biological Categorization and Reasoning Within and Across Cultures," *Cognition*, 84 (2002): 1–53; S. Gelman, *The Essential Child: Origins of Essentialism in Everyday Thought* (New York: Oxford University Press, 2003); L. Hirschfeld, *Race in the Making* (Cambridge, MA: MIT Press, 1996).

15. Aristotle, *Politics*, trans. Ernest Barker (New York: Oxford University Press, 1998), I.6,1254b, 27–39. There are also other places in which he compares slaves with nonhuman animals. See A. Pagden, *The Fall of Natural Man: The American Indian and the Origins of Comparative Ethnology* (Cambridge: Cambridge University Press, 1982) and B. Isaac, *The Invention of Racism in Classical Antiquity* (Princeton: Princeton University Press, 2004).

16. Aristotle, *Politics*, 1256b, 23–26. Cited and translated by B. Isaac, *The Invention of Racism in Classical Antiquity*, 178–179.

17. For an excellent, succinct discussion of Aristotle's theory of natural slavery see P. Garnsey, *Ideas of Slavery from Aristotle to Augustine* (Cambridge: Cambridge University Press, 1996). For a more technical treatment, see E. Garver, "Aristotle's Natural Slaves: Incomplete *Praxeis* and Incomplete Human Beings," *Journal of the History of Philosophy*, 32, no. 2 (1994): 173–195. For its transmission to medieval Islam, see P-A Hardy, "Medieval Muslim Philosophers on Race," in *Philosophers on Race: Critical Essays*, eds. J. K. Ward and T. L. Lott (Malden, MA: Blackwell, 2002).

18. Augustine, *The City of God*, 16.8.

19. Boethius, *The Consolation of Philosophy*, trans. W. V. Cooper (London: J. M. Dent, 1902), 113–114. For the use of animal metaphors in classical literature, see J. Gottschall, "Homer's Human Animal: Ritual Combat in the *Iliad*," *Philosophy and Literature* 25 (2001): 278–294 and K. Bradley, "Animalizing the Slave: The Truth of Fiction," *The Journal of Roman Studies* 90 (2000): 110–125.

20. A. Pope, *Essay on Man and Satires* (Teddington, UK: Echo Library, 1977), 8. For a twentieth-century defense of the great chain, see E. F. Schumacher, *A Guide for the Perplexed* (New York: Harper and Row, 1977).

21. A. O. Lovejoy, *The Great Chain of Being: The History of an Idea* (Cambridge, MA: Harvard University Press, 1936). K. W. Luckert, *Egyptian Light and Hebrew Fire: Theological and Philosophical Roots of Christendom in Evolutionary Perspective* (Albany: SUNY Press, 1991).

22. T. Jefferson, "Notes on the State of Virginia," *Political Writings* (Cambridge: Cambridge University Press, 1999). This weird idea had considerable currency during the seventeenth and eighteenth centuries. Jefferson may have picked it up from various sources, including Sir Thomas Herbert, who claimed that African women "keep company" with baboons and Voltaire's claim that "in the hot countries apes subjugated girls." T. Herbert, *A Relation of Some Yeares Travaile* (London: Jacob Blome and Richard Bishop, 1637), 19. F. M. Voltaire, "Of the Different Races of Men," in *The Idea of Race*, eds. R. Bernasconi and T. L. Lott (Indianapolis: Hackett, 2000), 6.

23. W. Shakespeare, *Othello*, 2.3: 262–264.

24. The Koran, 8:55 (trans. M. H. Shakir).
25. See also Sura 7:166: "When in their insolence they transgressed (all) prohibitions,We said to them: 'Be ye apes, despised and rejected.'"
26. *Sahih al-Bukhari*, Vol. 4, Book 54, No. 524; Vol. 7, Book 69, No. 494v; Vol. 4, Book 55, No. 569. Also *Sahih Muslim*, Book 042, Number 7135 and 7136; Book 033, Number 6438. The Koran contains references to various biblical figures, including Jesus, Mary, Abraham, Solomon, and David. Muslims believe that both Jewish and Christian scriptures were inspired by Allah.
27. A. Al-Azmeh, *Arabic Thought and Islamic Societies* (Dover, NH: Croom Helm, 1986).
28. P. della Mirandola, "Oration on the Dignity of Man," trans. R. Hooker, http://www.wsu.edu:8001/~dee/REN/ORATION.HTM
29. Quoted in R. C. Dales, "A Medieval View of Human Dignity," *Journal of the History of Ideas* 38, no. 4 (1977): 557–572.
30. W. R. Newman, *Promethean Ambitions: Alchemy and the Quest to Perfect Nature* (Chicago: University of Chicago Press, 2004), 217.
31. T. Tryon, "Friendly Advice to the Gentlemen-Planters of the East and West Indies," in *Carribeana: English Literature of the West Indies, 1657–1777*, ed., T. W. Krise (Chicago: University of Chicago Press, 1999), 62.
32. E. C. Mossner, *The Life of David Hume* (New York: Oxford University Press, 2001).
33. D. Hume, *An Enquiry Concerning the Principles of Morals*, ed. L. A. Selby-Bigge, 3rd ed. revised by P. H. Nidditch (Oxford: Clarendon Press, 1975), *University Press*, 1998), 173–175.
34. D. Hume, *A Treatise of Human Nature* (London: Penguin, 1985), 397–398.
35. J. D. Frank, "Prenuclear-Age Leaders and the Nuclear Arms Race," *American Journal of Orthopsychiatry* 52 (1982), 633.
36. J. Locke, *Two Treatises of Government* [2:11]. J. Tully, *An Approach to Political Philosophy: Locke in Contexts* (Cambridge: Cambridge University Press, 1993).
37. Hume, *Treatise*, 414, 492, 489. For the common point of view, see R. Cohen, "The Common Point of View in Hume's Ethics," *Philosophy and Phenomenological Research* 57, no. 4 (1997): 827–850, and K. Korsgaard, "The General Point of View: Love and Moral Approval in Hume's Ethics," *Hume Studies* 25 no. 1 & 2 (1999): 3–42.
38. D. Hume, *An Enquiry Concerning the Principles of Morals*, 88.
39. Ibid., 88–89, emphasis added.
40. D. Hume, *Treatise*, 273–274, 369. See also A. Waldow, "Hume's Belief in Other Minds," *British Journal for the History of Philosophy* 17, no. 1 (2009): 199–132.
41. D. Hume, *The Natural History of Religion* (Whitefish, MT: Kessinger, 2004), 8.

For more on the anthropomorphizing impulse, see S. Guthrie, *Faces in the Clouds: A New Theory of Religion* (New York: Oxford University Press, 1993).

42. For Hume's views on imagination, see G. Streminger, "Hume's Theory of Imagination," *Hume Studies* 6, no. 2 (1980): 98–118. For extended discussion of the passage in question, see A. Kuflick, "Hume on Justice to Animals, Indians and Women," *Hume Studies* 24, no. 1 (1998): 53–70 and D. M. Levy and S. Peart, "Sympathy and Approbation in Hume and Smith: A Solution to the Other Rational Species Problem," *Economics and Philosophy* 20 (2004): 331–349. For Hume on women, see *Enquiry*, 89.

43. D. Diderot, anonymous passage in G. T. F. Raynal, *L' Histoire philosophique et politique des établissements et du commerce des Européens dans les deux Indies* (1770), quoted in A. Fitzmaurice, "Anticolonialism in Western Political Thought: The Colonial Origins of the Concept of Genocide," in *Empire, Colony, Genocide: Conquest, Occupation, and Subaltern Resistance in World History*, ed. A. D. Moses (New York: Berghahn, 2008), 67.

44. L. W. Beck, *Early German Philosophy: Kant and His Predecessors* (Cambridge, MA: Harvard/Belknap, 1969).

45. I. Kant, *Grounding for the Metaphysics of Morals*, 3rd ed. trans. J. W. Ellington (Indianapolis, IN: Hackett, 1993), 30.

46. I. Kant, *Anthropology from a Pragmatic Point of View*, trans. M. J. Gregor (Berlin: Springer Verlag, 1974), 9, and *Grounding*, 428.

47. I. Kant, "Conjectures on the Beginning of Human History," in *Kant: Political Writings*, trans. H. B. Nesbit, ed. H. Reiss (Cambridge: Cambridge University Press, 1991), 125. For a nuanced discussion of Kant's views on nonhuman animals, see C. Korsgaard, "Fellow Creatures: Kantian Ethics and Our Duties to Animals," in *The Tanner Lectures on Human Values* 25/26, ed. G. B. Peterson (Salt Lake City, UT: University of Utah Press, 2005).

48. L. Doyle, "'A Dead Iraqi Is Just Another Dead Iraqi.' You Know, So What?," *The Independent*, July 12, 2007. Quoted in Steuter and Wills, *At War with Metaphor*, 85.

49. W. G. Sumner, *Folkways: The Study of Mores, Manners, Customs and Morals* (Mineola, NY: Dover, 2002), 13.

50. J. Diamond, *The Rise and Fall of the Third Chimpanzee* (London: Vintage, 1991), 267.

51. Sumner, *Folkways*, 14.

52. F. Boas, "Individual, Family, Population and Race," *Proceedings of the American Philosophical Society* 87, no 2. (1943): 161.

53. F. Roes, "An Interview with Napoleon Chagnon," *Human Ethology Bulletin* 13, no. 4, 1998: 6.

54. Sumner, *Folkways*, 14–15.

55. W. Owen, *The Collected Poems of Wilfred Owen* (New York: New Directions, 1965), 44.

56. H. Banister and O. L. Zangwill, "John Thomson MacCurdy, 1886–1947," *British Journal of Psychology* 40 (1949): 1–4; J. Forrester, "1919: Psychology and Psychoanalysis, Cambridge and London—Myers, Jones and MacCurdy," *Psychoanalysis and History* 10, no. 1 (2008): 37–94.

57. J. T. MacCurdy, *The Psychology of War* (Boston: John W. Luce and Company, 1918), 40.

58. Ibid., 41. See also D. L. Smith, *The Most Dangerous Animal: Human Nature and the Origins of War* (New York: St. Martin's Press, 2007).

59. MacCurdy, *The Psychology of War*, 38–39.

60. Ibid., 53.

61. L. Cohen, "The Story of Isaac," in *Stranger Music: The Poems and Songs of Leonard Cohen* (New York: Pantheon Books, 1993).

62. Erikson introduced the concept of pseudospeciation in "Ontogeny of Ritualization in Man," *Philosophical Transactions of the Royal Society of London*, series B, 251 (1966): 337–349. He discusses pseudospeciation in *Gandhi's Truth: On the Origin of Militant Nonviolence* (New York: Norton, 1994), *Identity, Youth and Crisis* (New York: Norton, 1968), and in "The Galilian Sayings and the Sense of 'I,'" in R. Coles (ed.), *The Erik Erikson Reader* (New York: Norton, 2001). His single paper on the subject was an address given at the American Psychiatric Association in May 1984 and published as "Pseudospeciation in the Nuclear Age," *Political Psychology* 6, no. 2 (1984): 213–217. For Lorenz's suggestion, see L. J. Friedman, *Identity's Architect: A Biography of Erik H. Erikson* (Cambridge, MA: Harvard University Press, 2000).

63. E. H. Erikson, "Pseudospeciation in the Nuclear Age," *Political Psychology* 6, no. 2 (1984): 214.

64. Ibid.

65. K. Lorenz, *On Aggression* (New York: Bantam, 1966), 79.

66. Quoted in B. Müller-Hill, *Murderous Science: Elimination by Scientific Selection of Jews, Gypsies and Others in Germany, 1933–1945* (Woodbury, NY: Cold Spring Harbor Laboratory Press, 1998), 14.

67. K. N. Laland and B. G. Galef (eds.), *The Question of Animal Culture* (Cambridge, MA: Harvard University Press, 2009).

68. R. Wrangham and D. Peterson, *Demonic Males: Apes and the Origin of Human Violence* (New York: Matiner, 1997), 8–9.

69. E. O. Wilson, *Sociobiology: The New Synthesis* (Cambridge, MA: Harvard, 1975). I. Eibl-Eibesfeldt, *The Biology of War and Peace: Men, Animals and Aggression* (New York: Viking, 1979).

70. E. O. Wilson, *On Human Nature* (Cambridge, MA: Harvard University Press, 1978), 70.

71. Eibl-Eibesfeldt, *The Biology of War and Peace*, 123.

72. Wrangham and Peterson, *Demonic Males*, 70.

73. J. Goodall, *Through a Window: My Thirty Years with the Chimpanzees of Gombe* (New York: Mariner, 2000), 209–210, emphasis added. See also P. Roscoe, "Intelligence, Coalitional Killing, and the Antecedents of War," *American Anthropologist* 109, no. 3 (2007): 485–495.

74. I. Eibl-Eibesfeldt, *Human Ethology* (New York: Aldine de Gruyter, 1989), 402.

3. CALIBAN'S CHILDREN

1. J-P. Sartre, preface to F. Fanon, *The Wretched of the Earth*, trans. C. Farrington (New York: Grove Weidenfeld, 1968), 26.

2. See A. Pagden, *The Fall of Natural Man: The American Indian and the Origins of Comparative Ethnology* (Cambridge: Cambridge University Press, 1982).

3. F. Retamar, "Caliban," *The Massachusetts Review* 15 (1973–74): 24.

4. Polybius, *The Complete Histories of Polybius*, trans. W. R. Paton (Lawrence, KS: Digireads.com, 2009), 48.

5. Quoted in S. W. Baron, *A Social and Religious History of the Jews*, vol. 8 (New York: Columbia University Press, 1970), 135.

6. H. Kamen, *The Spanish Inquisition: A Historical Review* (Newhaven, CT: Yale University Press, 1999). D. Root, "Speaking Christian: Orthodoxy and Difference in 16th Century Spain," *Representations* 23 (1988): 118–134. A. Majid, *We Are All Moors: Ending Centuries of Crusades Against Muslims and Other Minorities* (Minneapolis: University of Minnesota Press, 2009).

7. Bartholomew Senarega, quoted in D. E. Stannard, *American Holocaust* (New York: Oxford University Press, 1992), 62.

8. J. G. Varner and J. J. Varner, *Dogs of the Conquest* (Norman, OK: University of Oklahima Press, 1983).

9. B. de Las Casas, *History of the Indies*, trans. A. Collard (New York: Harper and Row, 1971), 94.

10. H. R. Parish (ed.), *Bartolomé de las Casas: The Only Way* (New York, NY: Paulist Press, 1992), 12.

11. Las Casas, *History of the Indies*,

12. Las Casas, quoted in T. Todorov, *The Conquest of America: The Question of the Other* (Norman, OK: University of Oklahoma Press, 1999), 141.

13. B. Kiernan, *Blood and Soil: A World History of Genocide and Extermination from Sparta to Darfur* (Newhaven, CT: Yale University Press, 2007).

14. Quoted in Todorov, 141.

15. Ibid., 139.

16. P. Bakewell, *A History of Latin America: Empires and Sequels 1450–1930* (Massachusetts: Blackwell, 1997), 83.

17. The remark from Mair's *In Secundum Sententiarum* is quoted in Pagden, *The Fall of Natural Man*, 38. Aristotle, *Physics*, 199a: 20–25, in T. Irwin and G. Fine, *Aristotle Selections* (Indianapolis, IN: Hackett, 1995). M. R. Johnson, *Aristotle on Teleology* (New York: Oxford University Press, 2005).

18. Tostado and pseudo-Thomas are quoted in W. R. Newman, *Promethean Ambitions: Alchemy and the Quest to Perfect Nature* (Chicago: University of Chicago Press, 2004), 188–189, 192–193.

19. Newman, *Promethean Ambitions*, 217, emphasis added.

20. Paracelsus, *De homunculis*, quoted in Newman, *Promethean Ambitions*, 219. As an interesting aside, Newman remarks (vis-à-vis Paracelsus's strange sexual preoccupations) that forensic examination of his skeleton suggests that "he" was either a pseudohermaphroditic genetic male, or a genetic female suffering from adrenogenital syndrome. For Paracelsus's sexuality, see Newman, 196–197.

21. J. G. de Sepúlveda, *Tratado sobre las Justas Causas de la Guerra contra los Indios*, trans. M. Menendez, P. Garcia-Pelayo, and M. Garcia-Pelayo (Mexico City: Fondo de Cultura Económica, 1941), 153. Todorov, 152; Stannard, *American Holocaust*, 210; Pagden, *The Fall of Natural Man*, 23, 104, 116, 118. H. Honour, *The New Golden Land: European Images of America from the Discoveries to the Present Time* (New York: Pantheon, 1975). See also Kiernan, *Blood and Soil*, 83 and P. Mason, *Deconstructing America: Representation of the Other* (New York: Routledge, 1990).

22. Stannard, *American Holocaust*, 211. Paul III: Sublimis Deus, June 2, 1537, in H. R. Parish and H. E. Weidman, *Las Casas en Mexico: Historia y obras desconocidas* (Mexico City: Fondo De Cultura Economica, 1992). E. F. Fischer, *Indigenous Peoples, Civil Society, and the Neo-Liberal State in Latin America* (Oxford: Berghahn Books, 2009).

23. G. Jahoda, *Trail of Tears* (New York: Random House, 1995), 135. Letter from Reverend Solomon Stoddard to Governor Joseph Dudley, in R. Demos, *Remarkable Providences: Readings in Early American History* (Lebanon, NH: University Press of New England, 1991), 273.

24. J. Smith, *Generall Historie of Virginia, New England and the Summer Isles* (1624), quoted in M. J. Bowden, "The Invention of American Tradition," *Journal of Historical Geography* 18, no. 1 1992:3–26. S. Purchas, 19, *Hakluytus Postumus or Purchas His Pilgrimes* 231 (1625), quoted in A. Tsesis, *Destructive Messages: How Hate Speech Paves the Way for Harmful Social Movements* (New York: New York

University Press, 2002). C. Brooke, *A Poem on the Late Massacre in Virginia, With Particular Mention of Those Men of Note That Suffered in That Disaster* (London: G. Eld for Robert Myldbourne, 1622), 22–23. See also F. F. Fausz, "The First Act of Terrorism in English America," *History News Network,* January 16, 2006 (http://hnn.us/articles/19085.html); A. T. Vaughan, *Roots of American Racism: Essays on the Colonial Experience* (Oxford: Oxford University Press, 1995).

25. W. Bradford, *Of Plymouth Plantation, 1620–1647* (New York: Modern Library, 1981), 296.

26. W. Winthrop, "Some Meditations" (1675), quoted in R. Drinnon, *Facing West: Metaphysics of Indian-Hating and Empire Building* (New York: Meridian, 1980), 54.

27. Quoted in Vaughan, *Roots of American Racism,* 24–25.

28. R. F. Berkhoffer, Jr., *The White Man's Indian: Images of the American Indian from Columbus to the Present* (New York: Vintage, 1979), 13. See also R. Bernheimer, *Wild Men in the Middle Ages: A Study in Art, Sentiment and Demonology* (Cambridge, MA: Harvard University Press, 1952). James I is quoted in Vaughan, *Roots of American Racism,* 12. Cotton Mather is quoted in Vaughan, page 24.

29. The letter from Washington is quoted in R. Horseman, "American Indian Policy in the Old Northwest, 1783–1812," in R. L. Nichols (ed.), *The American Indian: Past and Present* (New York: Random House, 1986), 139.

30. C. D. Eby, "That Disgraceful Affair," *The Black Hawk War* (New York: Norton, 1973), 259. H. H. Brackenridge, *Indian Atrocities* (New York: V. P. James, 1967), 62. The passage from *The Oregon Trail* is quoted in R. F. Berkhoffer, Jr., *The White Man's Indian: Images of the American Indian from Columbus to the Present* (New York: Vintage, 1979), 96.

31. D. E. Connor, *Joseph Reddeford Walker and the Arizona Adventure* (Norman, OK: University of Oklahoma Press, 1956), 302–303. J. R. Brown, *Adventures in the Apache Country: A Tour Through Arizona and Sonora* (New York: Harper and Brothers, 1869), 100. These and many other examples of subhuman characterizations of Apaches are cited in K. Jacoby, "'The Broad Platform of Extermination': Nature and Violence in Nineteenth Century North American Borderlands," *Journal of Genocide Research* 10, no. 2 (2008): 249–267.

32. Jacoby, "'The Broad Platform of Extermination,'" 258.

33. Haslam, "Dehumanization: An Integrative Review," *Personality and Social Psychology Review* 10, no. 3 (2006): 255.

34. See, for example, R. E. Green et al., "Analysis of One Million Base Pairs of Neanderthal DNA," *Nature* 444 (2006): 330–336; P. Forster, "Ice Ages and the Mitochondrial DNA Chronology of Human Dispersals: A Review," *Philosophical*

Transactions of The Royal Society of London: Biological Sciences 359, no. 1442 (2004): 255–264; B. Wood and M. Collard, "The human genus," *Science* 284 (1999): 64–71; M. Hasegawa, H. Kishino, and T. Yano, "Dating of the Human-Ape Splitting by a Molecular Clock of Mitochondrial DNA," *Journal of Molecular Evolution* 22, no. 2 (1985): 160–174; A. S. Ryan and D. C. Johansen, "Anterior Dental Microwear in *Australopithecus afarensis*: Comparisons with Human and Nonhuman Primates," *Journal of Human Evolution* 18, no. 3 (1989): 235–268.

35. D. C. Dennett, *Darwin's Dangerous Idea: Evolution and the Meanings of Life* (New York: Simon and Schuster, 1996).

36. H. C. Kelman, "Violence Without Moral Restraint: Reflections on the Dehumanization of Victims and Victimizers," *Journal of Social Issues* 29, no. 4 (1973): 24.

37. Ibid., 48–49.

38. Ibid., 49.

39. L. Kuper, *Genocide: Its Political Use in the Twentieth Century* (Newhaven, CT: Yale University Press, 1981), 87.

40. Sura 9:29. For a fascinating discussion of the historically close relationship between Muslims and Jews, see Majid, *We Are All Moors*.

41. Al Ghazali, Kitab al-Wagiz fi Fiqh madhab al-imam al Safi'i. Quoted in A. G. Bostom (ed.), *The Legacy of Jihad: Islamic Holy War and the Fate of Non-Muslims* (London: Prometheus, 2008), 199.

42. W. I. Brustein, *Roots of Hate: Anti-Semitism in Europe Before the Holocaust* (Cambridge: Cambridge University Press, 2003).

43. C. Koonz, *The Nazi Conscience* (Cambridge, MA: Harvard University Press, 2003), 173.

44. G. Orwell, "Marrakech," in *George Orwell: An Age Like This, 1920–1940* (Boston: David R. Godine, 2000), 388. E. Steuter and D. Wills, *At War With Metaphor: Media Propaganda and Racism in the War on Terror* (New York: Lexington, 2009), 27. O. Santa Ana, *Brown Tide Rising: Metaphors of Latinos in Contemporary American Public Discourse* (Austin: University of Texas Press, 2002), 72. L. Stoddard, *The Rising Tide of Color Against White World-Supremacy* (New York: Charles Scribner's Sons, 1922).

45. Quoted in C. Lomnitz and Rafael Sánchez, "United by Hate: The Uses of Anti-Semitism in Chávez's Venezuela," *The Boston Review*, July–August 2009, emphasis added. A. Barrioneuvo, "Inquiry on 1994 Blast at Argentina Jewish Center Gets New Life," *New York Times*, July 17, 2009.

46. P. J. Oakes et al., "Becoming an In-group: Reexamining the Impact of Familiarity on Perceptions of Group Homogeneity," *Social Psychology Quarterly* 58, no. 1 (1995): 52-51. G. W. Allport, *The Nature of Prejudice* (Cambridge, MA: Addison-Wesley, 1954).

47. M. Heidegger, "Bekenntniss der Professoren," quoted in C. Koonz, *The Nazi Conscience*, 46.

48. N. Haslam, et. al., "Attributing and Denying Humanness to Others," *European Journal of Social Psychology* 19 (2008): 58. Plato, *Statesman*, 267E. The anecdote about Diogenes of Sinope is from Diogenes Laertius's *Lives of Eminent Philosophers*, 6.40.

49. See E. Ben-Ari, *Mastering Soldiers: Conflict, Emotions and the Enemy in an Israeli Military Unit* (New York: Berghahn Books, 1998).

50. M. Sendivogius, "The New Chemical Light," in A. E. Waite (ed.), *The Hermetic Museum: Containing Twenty-Two Most Celebrated Chemical Tracts* (London: James Elliot & Co., 1893), 106.

51. J. W. Yolton (ed.), *The Locke Reader* (Cambridge: Cambridge University Press, 1977), 55.

52. J. Locke, *An Essay Concerning Human Understanding* (London: William Tegg & Company, 1879), 518.

53. Ibid., 302.

54. See R. A. Wilson, M. N. Barker, and I. Brigant, "When Traditional Essentialism Fails: Biological Natural Kinds," *Philosophical Topics* 35 (102): 189–215.

55. S. Kripke, *Naming and Necessity* (Malden, MA: Blackwell, 1972), 123–125. Given the sequence of themes here, I want to make it clear that I am not endorsing (or rejecting) John Mackie's reading of Locke as a precursor of Kripke. See J. L. Mackie, "Locke's Anticipation of Kripke," *Analysis* 34 (1974): 177–180.

56. See D. Chalmers, *The Conscious Mind: In Search of a Fundamental Theory* (New York: Oxford University Press, 1996) and, for critique, D. Dennett, "The Unimagined Preposterousness of Zombies," *Journal of Consciousness Studies* 2, no. 4 (1995): 322–326.

4. THE RHETORIC OF ENMITY

1. N. J. O'Shaughnessy, *Politics and Propaganda: Weapons of Mass Seduction* (Ann Arbor: University of Michigan Press, 2004), 110.

2. E. Cassirer, *The Myth of the State* (Newhaven, CT: Yale University Press, 2009), 275.

3. Quoted in D. J. Goldhagen, *Worse Than War: Genocide, Eliminationism and the Ongoing Assault on Humanity* (New York: PublicAffairs, 2009), 158.

4. K. Bales, *Disposable People: The New Slavery and the Global Economy* (Berkeley: University of California Press, 1999), 197.

5. F. L. Olmsted, *The Cotton Kingdom: A Traveller's Observations on Cotton and Slavery in the American Slave States* 2 (New York: Mason Brothers, 1862), 203.

6. Ibid., 205–206.

7. Ephesians, 5: 6.

8. T. Wiedemann, *Greek and Roman Slavery* (London: Routledge,1981). This included the sexual enslavement of women. See G. Lerner, *The Creation of Patriarchy* (New York: Oxford University Press, 1986). For the origin of the term *servant*, see Florintinus, *Digesta seu Pandectae*, 1.5.4.2.

9. K. Jacoby, "Slaves by Nature? Domestic Animals and Human Slaves," *Slavery and Abolition: A Journal of Slave and Post-Slave Studies* 15, no. 1 [1994], 92–94. For the sacrifice of slaves, see M. A. Green, *Dying for the Gods: Human Sacrifice in Iron Age and Roman Europe* (Charleston, SC: Tempus Publishing, 2001).

10. K. Jacoby, "Slaves by Nature?"

11. Strabo, *The Geographies of Strabo*, trans. H. L. Jones (Cambridge, MA: Harvard university Press, 1917).

12. J. Heath, *The Talking Greeks: Speech, Animals, and the Other in Homer, Aeschylus, and Plato* (Cambridge: Cambridge University Press, 2005), 201.

13. L. Lowenthal and N. Guterman, *Prophets of Deceit: A Study of the Techniques of the American Agitator* (New York: Harper and Brothers, 1949), 80. G. L. Mosse, *Toward the Final Solution: A History of European Racism* (New York: Howard Fertig, Inc., 1978). T. J. Curran, *Xenophobia and Immigration, 1820–1930* (Boston: Thwayne Publishers, 1975).

14. M. Lichtheim, *Ancient Egyptian Literature, Vol. II: The New Kingdom* (Berkeley, CA: University of California Press, 1976), 144.

15. M-C. Poo, *Enemies of Civilization: Attitudes Towards Foreigners in Ancient Mesopotamia, Egypt and China* (Albany, NY: State University of New York Press, 2005), 74.

16. Ibid., 50–51.

17. M. H. Fried, *The Notion of Tribe* (Menlo Park: Cummings, 1975). F. Dikötter, *The Discourse of Race in Modern China* (Stanford, CA: Stanford University Press, 1992). Poo, *Enemies of Civilization*, 65. E. G. Pulleyblank, "The Origin and Nature of Chattel Slavery in China," *Journal of Economic and Social History of the Orient* 1 (1958): 209.

18. O. Patterson, *Slavery and Social Death: A. Comparative Study* (Cambridge, MA: Harvard University Press, 1985), 24.

19. Ibid., 25–26. See "So You Want to Own a Ball Club," *Forbes*, April 1, 1977, 37, cited in D. S. Eitzen and G. H. Sage, *The Sociology of American Sports* (Dubuque, IA: William C. Brown Company, 1977), 188.

20. Patterson *Slavery and Social Death*, 47.

21. K. Bradley, "Animalizing the Slave: The Truth of Fiction," *The Journal of Roman Studies* 90 (2000): 110. See also F. D. Harvey, "Herodotus and the Man-Footed

Creature," in L. Archer (ed.), *Slavery and Other Forms of Unfree Labor* (New York: Routledge, 1988).

22. P. Garnsey, *Ideas of Slavery from Aristotle to Augustine* (Cambridge: Cambridge University Press, 1996).

23. Bradley, "Animalizing the Slave," 111. M. Gordon, *Slavery in the Arab World* (New York: New Amsterdam Books, 1989).

24. F. Douglass, *Narrative of the Life of Frederick Douglass, an American Slave* (New York: Barnes and Noble Books, 2005), 49.

25. A. de Tocqueville, *Democracy in America*, trans. George Lawrence (New York: Harper Perennial Modern Classics, 2000), 32.

26. E. Hornung, *The Ancient Egyptian Book of the Afterlife* (Ithaca: Cornell University Press, 1999). R. Parkinson, *The Tale of Sinuhe and Other Ancient Egyptian Poems* (Oxford: Oxford University Press, 1997). For differences in skin color as a racial marker, see E. Bresciani, "Foreigners," in S. Donadoni (ed.), *The Egyptians* (Chicago: University of Chicago Press, 1997). Galen's remark is from a work that has not survived in the original Greek, but is quoted in al-Ma'sūdī's *Mūruj al-Dhahab wa Ma'ādin al Jawhar*, quoted in English translation in R. Segal, *Islam's Black Slaves: The Other Black Diaspora* (New York: Farrar, Straus & Giroux, 2001), 46.

27. Patterson, *Slavery and Social Death*, 176. See F. M. Snowden, *Blacks in Antiquity: Ethiopians in the Greco-Roman Experience* (Cambridge, MA: Harvard University Press, 1971) and F. M. Snowden, *Before Color Prejudice: The Ancient View of Blacks* (Cambridge, MA: Harvard University Press, 1991).

28. The *hadith* about Muhammed's final sermon is found in the authoritative collections of both al-Bukhari and Muslim. Subhaym is quoted in B. Lewis, *Race and Slavery in the Middle East: An Historical Enquiry* (New York: Oxford University Press, 1990), 28. Al-Andalusi is quoted in Segal, *Islam's Black Slaves*, 49. The statement by Ibn Khaldun is from *The Muqaddimah*, vol. 1, trans. F. Rosenthal (New York: Pantheon, 1958), 301.

29. Quoted in M. Gordon, 102. See R. Segal, *Islam's Black Slaves*.

30. Patterson, *Slavery and Social Death*, 146.

31. W. J. Jordan, *White Over Black: American Attitudes Towards the Negro, 1550–1812* (Baltimore: Benguin, 1969), 228–229.

32. Information about Godwyn's life and quotations from his works are taken from A. T. Vaughan, "The Slaveholders' 'Hellish Principles': A Seventeenth Century Critique," Vaughan, *Roots of American Racism: Essays on the Colonial Experience* (Oxford: Oxford University Press, 1995), 55–81.

33. W. Updike, *A History of the Episcopal Church in Narragansett, Rhode Island: Including a History of the Other Episcopal Churches in the State* (Ithaca: Cornell

University Library, 2009), 211. E. G. Wood, *The Arrogance of Faith: Christianity and Race in America from the Colonial Era to the Twentieth Century* (New York: Alfred A. Knopf, 1900), 237. Jordan, *White Over Black*, 228. A. Tsesis, *Destructive Messages: How Hate Speech Paves the Way for Harmful Social Movements* (New York: New York University Press, 2002), 31–35. M. Godwyn, *The Negro's and Indians Advocate*, quoted in M. Cantor, "The Image of the Negro in Colonial Literature," *New England Quarterly* 36 (1963): 452–477. A. C. Fraser, *Life and Letters of George Berkeley, D.D., Formerly Bishop of Cloyne, Vol 4*. (London: Macmillan & Co, 1871), 188.

34. Jordan, *White Over Black*, 232.

35. For a similar view, see Daniel C. Dennett's remarks about "believing in belief" in his *Breaking the Spell: Religion as a Natural Phenomenon* (New York: Penguin, 2007) and the distinction between "opinion" and "belief" in his *Brainstorms: Philosophical Essays on Mind and Psychology* (Cambridge, MA: MIT Press, 1981).

36. Genesis, 9:19–27.

37. D. M. Goldenberg, *The Curse of Ham: Race and Slavery in Early Judaism, Christianity and Islam* (Princeton, NJ: Princeton University Press, 2003). For Son of Ham shows, see M. W. Robbins and W. Palitz, *Brooklyn: A State of Mind* (New York: Workman Publishing Company, 2001). For a good fictionalized account, see K. Baker, *Dreamland* (New York: HarperCollins, 1999).

38. Quoted in Vaughan, *Roots of American Racism*, 67.

39. Ibid., 66–67.

40. T. Waitz, *Introduction to Anthropology*, trans. J. Frederick Collingwood (London: Anthropological Society, 1863), 13.

41. Ibid., 351.

42. See note 43.

43. Lawrence is quoted in T. F. Gossett, *Race: The History of an Idea in America* (New York: Schoken, 1965), 57. For Darwin's views on race and slavery, see A. Desmond & J. Moore, *Darwin's Sacred Cause: How a Hatred of Slavery Shaped Darwin's Views on Human Evolution* (New York: Houghton Mifflin Harcourt, 2009). Nineteenth-century racist writings often had almost unbelievably wordy titles. B. H. Payne, *The Negro: What Is His Ethnological Status? Is He Progeny of Ham? Is He a Descendant of Adam and Eve? Has He a Soul? Or Is He a Beast in God's Nomenclature? What Is His Status as Fixed by God In Creation?* (Cincinnati, 1867). C. Carroll, "*The Negro A Beast*"; or, "*In the Image of God*"; *The Reasoner of the Age, the Revelator of the Century! The Bible as it Is! The Negro and His Relation to the Human Family! The Negro a Beast, but Created with Articulate Speech, and Hands, That He May Be of Service to His Master—the White*

Man! The Negro Not the Son of Ham, Neither Can It Be Proven by the Bible and the Argument of the Theologian Who Would Claim Such, Melts to Mist before the Thunderous and Convincing Arguments of this Masterful Book (St. Louis, 1900). For an excellent discussion of this genre, see E. G. Wood, *The Arrogance of Faith: Christianity and Race in America from the Colonial Era to the Twentieth Century* (New York: Alfred A. Knopf, 1900). For violence against African Americans in the late nineteenth and early twentieth centuries, see D. A. Blackmon, *Slavery by Another Name: The Re-Enslavement of Black Americans from the Civil War to World War II* (New York: Doubleday, 2008), E. Jaspin, *Buried in the Bitter Waters: The Hidden History of Racial Cleansing in America* (New York: Basic Books, 2007), and H. A. Washington, *Medical Apartheid: The Dark History of Medical Experimentation on Black Americans from Colonial Times to the Present* (New York: Doubleday, 2006). For a superb account of the pre-Adamite theory of race, and its connection with the monogenecist/polygenecist controversy, see David N. Livingstone's magisterial *Adam's Ancestors: Race, Religion and the Politics of Human Origins* (Baltimore: Johns Hopkins University Press, 2008).

44. Quoted in A. Hochschild, *King Leopold's Ghost: A Story of Greed, Terror and Heroism in Colonial Africa* (New York: Mariner, 1999), 166.

45. M. Keller, "The Scandal in the Zoo," *New York Times*, August 6, 2006. See also P. V. Bradford and H. Blume, *Ota Benga: The Pygmy in the Zoo* (New York: Delta, 1992).

46. Quoted in Bradford and Blume, *Ota Benga*, 182.

47. "Pygmy Chased by Crowd," *New York Times*, September 9, 1906.

48. D. Apel, *Imagery of Lynching: Black Men, White Women and the Mob* (New York: W. W. Norton & Company, 2003), 32.

49. A. Smith, *The Theory of Moral Sentiments* (Eastford, CT: Martino, 2009), 1.

50. A. C. Grayling, "Paradox at the Heart of Our Warring Psyche," *The Australian*, February 23, 2008.

51. H. Arendt, *Eichmann in Jerusalem: A Report on the Banality of Evil* (New York: Penguin, 1994).

52. S. Milgram, *Obedience to Authority* (New York: Harper & Row, 1974), 6.

53. "TRIALS: My Lai: A Question of Orders," *Time*, January 25, 1971.

54. H. C. Kelman, "Violence without Moral Restraint."

55. Ibid., 48. See also H. Fein, *Accounting for Genocide* (New York: Free Press, 1979), 30.

56. A. Bandura, B. Underwood, and M. E. Fromson, "Disinhibition of Aggression Through Diffusion of Responsibility and Dehumanization of Victims," *Journal of Research in Personality* 9 (1975), 266.

57. Ibid.

5. LEARNING FROM GENOCIDE

1. E. Litvinoff, "To T. S. Eliot," in H. Schwartz and A. Rudolf (eds.), *Voices Within the Ark* (New York: Avon, 1980), 715–716.

2. D. J. Goldhagen, *Worse Than War: Genocide, Eliminationism and the Ongoing Assault on Humanity* (New York: Public Affairs, 2009), 191–192.

3. D. D. Gilmore, *Monsters: Evil Beings, Mythical Beasts and All Manner of Imaginary Terrors* (Pittsburgh: University of Pennsylvania Press, 2003), 25.

4. Ibid., 31.

5. Ibid., 45. For an informative discussion of the monstrous races in relation to the colonial project, see P. Mason, *Deconstructing America: Representations of the Other* (New York: Routledge, 1990).

6. Pliny, *Natural History*, Book II, quoted in S. T. Asma, *On Monsters: An Unnatural History of Our Worst Fears* (New York: Oxford University Press, 2009), 33.

7. D. Gilmore, 65.

8. D. L. Jeffrey, "Medieval Monsters," in *Manlike Monsters on Trial: Early Records and Modern Evidence*, eds. M. Halpern and M. M. Ames (Vancouver: University of British Columbia Press, 1980). Quoted in Gilmore, *Worse Than War*, 60.

9. S. Drakulić, *They Would Never Hurt a Fly: War Criminals on Trial in the Hague* (London: Abacus, 2004), 168.

10. L. Goldensohn, *The Nuremberg Interviews* (New York: Vintage, 2005), quoted in S. K. Baum, *The Psychology of Genocide: Perpetrators, Bystanders and Rescuers* (Cambridge: Cambridge University Press, 2008), 74.

11. M. Sageman, *Understanding Terrorist Networks* (Philadelphia: University of Pennsylvania Press, 2004), 135.

12. F. Kafka, *The Metamorphosis* (Whitefish, MT: Kessinger, 2004), 1.

13. Quoted in S. Hornshøj-Møller & D. Cuthbert, " 'Der ewige Jude'(1940): Joseph Goebbels' Unequalled Monument to Anti-Semitism," *Historical Journal of Film, Radio and Television* 12, no. 1, 42.

14. J. Goebbels, *The Goebbels Diaries, 1939–1941*, transl. and ed. Fred Taylor (New York: G. P. Putnam's Sons, 1983), 23.

15. Quoted from Elke Fröhlich's German edition of the Goebbels diaries by S. Hornshøj-Møller, "The Eternal Jew—A blueprint for genocide in the Nazi film archives," working paper GS 05, Yale Center for International and Area Studies, 6. This is a slightly different translation than that appearing in Goebbels, 1983, 36. B. Sax, *Animals in the Third Reich: Pets, Scapegoats and the Holocaust* (New York: Continuum, 2002).

16. Quoted in Goldhagen, *Worse Than War*, 386.

17. In the German:

> Wo Ratten auch auftauchen, tragen sie Vernichtung ins Land, zerstören sie menschliche Güter und Nahrungsmittel. Auf diese Weise verbreiten sie (die Ratten) Krankheiten, Pest, Lepra, Typhus, Cholera, Ruhr u.s.w. Sie sind hinterlistig, feige und grausam und treten meist in grossen Scharen auf. Sie stellen unter den Tieren das Element der heimtückischen, unterirdischen Zerstörung dar, nicht anders als die Juden unter den Menschen.

18. I. Kershaw, *Hitler: 1889–1936, Hubris* (New York: W. W. Norton & Co., 2000), 66.

19. J. Y. Conen, *The Roots of Nazi Psychology: Hitler's Utopian Barbarism* (Lexington, KY: University of Kentucky Press, 2000), 29–30.

20. H. Himmler, "From a speech by Himmler before senior SS officers in Poznan, October 4th, 1943," in *Encyclopedia of Genocide, Vol 1*, ed. I. W. Charny (Washington, DC: ABC-CLIO, 1999), 241.

21. O. Bartov, *Hitler's Army: Soldiers, Nazis and War in the Third Reich* (New York: Oxford University Press, 1992), 116, 127.

22. R. Campbell, "Autumn," in *The Oxford Book of Twentieth Century English Verse*, ed. P. Larkin (New York: Oxford University Press, 1973), 336.

23. Goldhagen, *Worse Than War*, 319.

24. Stanton's analysis is available at the website for Genocide Watch at http://www.genocidewatch.org/8stages.htm. D. Moshman, "Us and Them: Identity and Genocide," *Identity: An International Journal of Theory and Research* 7, 2007: 121. See also, for example, L. M. Woolf and M. R. Hulsizer, "Psychosocial Roots of Genocide: Risk, Prevention and Intervention," *Journal of Genocide Research* 7, no. 1, 2005: 101–128, N. J. Kressel, *Mass Hate: The Global Rise of Genocide and Terror* (New York: Plenum, 1996), H. Hirsch, *Genocide and the Politics of Memory: Studying Death to Preserve Life* (Chapel Hill, NC: University of North Carolina Press, 1995), and D. Chirot and C. McCauley, *Why Not Kill Them All? The Logic and Prevention of Mass Political Murder* (Princeton, NJ: Princeton University Press, 2006), 80.

25. A. Hochschild, *King Leopold's Ghost: A Story of Greed, Terror and Heroism in Colonial Africa* (New York: Houghton Mifflin, 1999), 31.

26. H. Bley, *South-West Africa Under German Rule 1894–1914* (Evanston, Ill: Northwestern University Press, 1971): 97. Quoted in M. Mann, *The Dark Side of Democracy: Explaining Ethnic Cleansing* (New York: Cambridge University Press, 2005).

27. H. Dreschler, *"Let Us Die Fighting": The Struggle of the Herrero and the Nama Against German Imperialism* (London: Zed, 1980): 167–8, n. 6. Quoted in M. Mann, *The Dark Side of Democracy: Explaining Ethnic Cleansing* (New York: Cambridge University Press, 2005).

28. H. Dreschler, *"Let Us Die Fighting,"* 154. H. Bley, *South-West Africa Under German Rule 1894–1914,* 97. D. J. Schaller, "From Conquest to Genocide: Colonial rule in German Southwest Africa and German East Africa," in D. Moses, *Empire, Colony, Genocide: Conquest, Occupation and Subaltern Resistance in World History* (Oxford: Berghahn Books, 2008). The account of burning is from *Report on the Natives of South-West Africa and Their Treatment by Germany,* quoted in Goldhagen, *Worse Than War,* 181.

29. W. F. Ramsay, *Impressions of Turkey During Twelve Years' Wanderings* (New York: Hodder & Stoughton, 1897), 206–207.

30. V. N. Dadrian, *The History of the Armenian Genocide: Ethnic Conflict from the Balkans to Anatolia to the Caucasus* (New York: Berghahn Books, 1996), 159. The letter is quoted in Kiernan, 406.

31. V. Dadrian, "The Armenian Genocide: An Interpretation," in *America and the Armenian Genocide of 1915,* ed. J. Winter (New York: Cambridge University Press, 2004). V. Dadrian, *The History of the Armenian Genocide: Ethnic Conflict from the Balkans to Anatolia to the Caucasus* (New York: Berghahan Books, 2004). V. Dadrian, "The Role of Turkish Physicians in the World War I Genocide of Ottoman Armenians," *Holocaust and Genocide Studies* 1(1986): 175. M. Mann, *The Dark Side of Democracy: Explaining Ethnic Cleansing* (Cambridge: Cambridge University Press, 2004), 160. T. Ackam, *A Shameful Act: The Armenian Genocide and the Question of Turkish Responsibility* (New York: Macmillan, 2007). Y. Aurun, "The Holocaust: Responses of the Jewish Community in Palestine," in *Encyclopedia of Genocide,* ed. I. W. Charny (Oxford: ABC-CLIO, 1999). J. Bourke, "Barbarization vs Civilization in Time of War," in *The Barbarization of Warfare,* ed. G. Kassimeris (New York: New York Universities Press, 2006).

32. "Memorandum des SD-Amtes," to Heydrich, quoted in C. Koonz, *The Nazi Conscience* (Cambridge, MA: Harvard/Belknap, 2003), 246. N. H. Baynes, *Hitler's Speeches* (London: Oxford University Press, 1942).

33. A. Musolff, "What Role Do Metaphors Play in Racial Prejudice? The Function of Antisemitic Imagery in Hitler's *Mein Kampf,*" *Patterns of Prejudice* 41, no. 1 (2007): 25.

34. Ibid. T. Mommsen, *History of Rome* 3 (Ithaca, NY: Cornell University Press, 2009).

35. Quoted in E. Jäckel, *Hitler's World View: A Blueprint for Power* (Cambridge, MA: Harvard University Press, 1981), 89.

36. Quoted in Musolff, "What Role Do Metaphors Play in Racial Prejudice," 31, 33.

37. V. E. Bonnell, *Iconography of Power: Soviet Political Posters Under Lenin and Stalin* (Berkeley, CA: University of California Press, 1998). A. Nove, "Victims of

Stalinism, How Many?" in *Stalinist Terror: New Perspectives*, eds. J. A. Getty and R. T. Manning (Cambridge: Cambridge University Press, 1993).

38. V. Grossman, *Forever Flowing* (Chicago: Northwestern University Press, 1997), 142–143.

39. Ibid., 144.

40. X. Peng, "Demographic Consequences of the Great Leap Forward in China's Provinces," *Population and Development Review* 13, no. 4 (1987): 639–670.

41. X. Lu, *The Rhetoric of the Chinese Cultural Revolution: The Impact on Chinese Thought, Culture and Communication* (Columbia, SC: University of South Carolina Press, 2004).

42. Ibid., 60.

43. Z. P. Luo, *A Generation Lost: China Under the Cultural Revolution* (New York: Henry Holt, 1990), 28. Quoted in X. Lu, *The Rhetoric of the Chinese Cultural Revolution*, 92.

44. B. Kiernan, "The Cambodian Genocide," in *Encyclopedia of Genocide*, vol. 1, ed. I. Charny (Santa Barbara, CA: ABC-CLIO, 1999), 131.

45. B. Kiernan, *Blood and Soil: A World History of Genocide and Extermination from Sparta to Darfur* (Newhaven, CT: Yale University Press, 2007), 549.

46. B. Kiernan, "The Cambodian Genocide." P. Yathay, *Stay Alive, My Son* (New York: Touchstone, 1987), 73.

47. Quoted in Kiernan, *Blood and Soil*, 549–550.

48. N. Moyer, *Escape from the Killing Fields* (Grand Rapids, MI: Zonervan Publishing House, 1991), 123. A. L. Hinton, "Agents of Death: Explaining the Cambodian Genocide in Terms of Psychosocial Dissonance," *American Anthropologist* 98, no. 4 (1996): 818–831.

49. J. D. Criddle and T. B. Mam, *To Destroy You Is No Loss* (New York: Anchor, 1987), 164. L. Picq, *Beyond the Horizon: Five Years with the Khmer Rouge* (New York: St. Martin's Press), 100. D. Chandler, B. Kiernan and C. Boua, *Pol Pot Plans the Future: Confidential Leadership Documents from Democratic Campuchea, 1976–1977*. Monograph Series, 33 (New Haven CT: Yale University Southeast Asia Studies, 1988), 183. M. Stuart-Fox, *The Murderous Revolution* (Chippendale, Australia: Alternative Publishing Cooperative, 1985). A. L. Hinton, "Comrade Ox Did Not Object When His Family Was Killed," in I. W. Charny, *Encyclopedia of Genocide, Vol. I.*, 135. D. Chandler, *Voices from S-21: Terror and History in Pol Pot's Secret Prison* (Berkeley, CA: University of California Press, 1999). Goldhagen, *Worse Than War*, 371.

50. S. Mouth, "Imprinting Compassion," in *Children of Cambodia's Killing Fields: Memoirs by Survivors*, ed. K. DePaul (Newhaven, CT: Yale University Press, 1997), 179–180.

51. Quoted in Desforges, *Leave None to Tell the Story: Genocide in Rwanda* (Washington, DC: Human Rights Watch, 1999), 73.
52. B. Kiernan, *Blood and Soil,* 559.
53. The "cockroach" epithet became denigrating only secondarily. Originally, Tutsi militias referred to themselves as *inyenzi,* which was an acronym for a phrase roughly translated as "an insurgent who has committed himself to bravery." The cockroach is a common choice for representing traditional enemies. When the eighteenth century Danish biologist Linnaeus gave the most common species of this insect a scientific name, he chose *Blattella germanica* (roughly, "German cockroach") for no apparent reason other than hostility toward Germans. In Northern Germany the insects were called *Schwabe* ("Swabians")—a derogatory reference to southern Germans. Not to be outdone, southern Germans referred to them as *Preusse* ("Prussian"). In western Germany they were *Franzose* ("French") and in eastern Germany they were *Russe* ("Russians"). In Poland it's *prusak,* which means "Prussian," and in Newfoundland, they are "Yankee settlers." M. Berenbaum, "Freedom Roaches," *American Entomologist* 51, no. 1 (2005): 4,5, 10. J. Waller, *Becoming Evil: How Ordinary People Commit Genocide and Mass Killing* (Oxford: Oxford University Press, 2002).
54. Desforges, *Leave None to Tell the Story.*
55. R. Block, "The tragedy of Ruanda," *New York Review,* October 20, 1994. Kiernan, *Blood and Soil.* C. Kagwi-Ndungu, *The Challenges in Prosecuting Print Media for Incitement to Genocide,* International Development Research Center. http://www.idrc.ca/fr/ev-108292-201-1-DO_TOPIC.html The comment by Rakiya Omaar is from M. Montgomery and S. Smith, "The Few Who Stayed: Defying Genocide in Rwanda," *American Radioworks,* http://americanradioworks.publicradio.org/features/rwanda/segc2.html. J. Waller, *Becoming Evil: How Ordinary People Commit Genocide and Mass Killing* (Oxford: Oxford University Press, 2002), 247. J. M. V. Higiro, "Rwandan Private Print Media on the Eve of the Genocide," in *The Media and the Rwanda Genocide,* ed. A. Thompson (London: Pluto Press, 2007). B. Nowrojee, "A Lost Opportunity for Justice: Why Did the ICTR Not Prosecute Gender Propaganda?," in *The Media and the Rwanda Genocide,* ed. E. Thompson (London: Pluto Press, 2007). For "Operation Insecticide," see A. Desforges, *Leave None to Tell the Story.*
56. Goldhagen, *Worse Than War,* 353.
57. Ibid., 182.
58. G. Prunier, *Darfur: The Ambiguous Genocide* (Ithaca, NY: Cornell University Press, 2005).
59. J. Hagan and W. Rymond-Richmond, "The Collective Dynamics of Racial Dehumanization and Genocidal Victimization in Darfur," *American Sociological*

Review 73: 875–902. Mahmood Mamdani argues that ethic conflict in Darfur is of recent vintage; see M. Mamdani, *Saviors and Survivors: Darfur, Politics and the War on Terror* (New York: Pantheon, 2009).

60. Hagan and Rymond-Richmond, "The Collective Dynamics of Racial Dehumanization and Genocidal Victimization in Darfur," 882. H. Bashir, *Tears of the Desert: A Memoir of Survival in Darfur* (New York: Random House, 2009), 240.

61. W. Shakespeare, *Romeo and Juliet*, Act 3, Scene 3: 108–112.

62. This and all subsequent quotes from *The Subhuman* are from the English translation of *Der Untermensch* by Hermann Feuer at <http://www.holocaustre searchproject.org/holoprelude/deruntermensch.html>

63. Charny, *Encyclopedia of Genocide*, 241.

64. *Der Untermensch*, trans. Hermann Feuer.

65. L. Rees, *Auschwitz: The Nazis and the Final Solution* (London: BBC Books, 2001), 139.

66. The quotation and poem are from C. Koonz, *The Nazi Conscience* (Cambridge, MA: Harvard University Press, 2003), 137.

67. Ibid., 116.

68. Ibid., 197.

69. Ibid.

70. H. Trevor-Roper and A. Francois-Poncet (eds.) *Hitler's Politisches Testament: Die Bormann Diktate vom Februar und April 1945* (Hamburg: Albrecht Knaus, 1981), 66–69. Quoted in English translation in G. Heinsohn, "What Makes the Holocaust a Uniquely Unique Genocide?," *Journal of Genocide Research* 2, no. 3 (2000), 412.

71. Koonz, *The Nazi Conscience*, Chapter 7.

6. RACE

1. L. E. Smith, *Killers of the Dream* (New York: W. W. Norton & Company, 1994), 13.

2. Ibid., 13–14.

3. B. Latter, "Genetic Differences Within and Between Populations of the Major Human Subgroups," *American Naturalist* 116 (1980): 220–237, R. Lewontin, "The Aportionment of Human Diversity," *Evolutionary Biology* 25 (1972): 276–280, R. Lewontin, "Are the Races Different?," in D. Gill and L. Levidow (eds.), *Anti-Racist Science Teaching* (London: Free Association Books, 1987). For a fine critical discussion of the significance of race for contemporary biology, see S. M. Fullerton, "On the Absence of Biology in Philosophical Considerations of Race," in *Race and the Epistemologies of Ignorance*, eds. S. Sullivan and N. Tuana

(Albany: State University of New York Press, 2007) and also P. Kitcher, "Race, Ethnicity, Biology, Culture," in *In Mendel's Mirror: Philosophical Reflections on Biology* (New York: Oxford University Press, 2002). T. H. Huxley, A. C. Haddon, and A. Carr-Saunders, *We Europeans* (London: Jonathan Cape, 1935), 266–267.

4. L. E. Smith, *Killers of the Dream*, 35–38.

5. For an excellent discussion of the debate between racial skeptics and racial constructionists, see R. Mallon, "Passing, traveling and reality: social constructionism and the metaphysics of race," *Noûs* 38, no. 4 [2004]: 644–673. For examples of constructionist theories, see G. M. Fredrickson, *The Arrogance of Race: Historical Perspectives on Slavery, Racism and Social Equality* (Middletown, CT: Wesleyan University Press, 1988); A. Smedley, *Race in North America: Origin and Evolution of a Worldview* (Boulder, CO: Westview Press, 1993).

6. To appreciate the full horror of life for African-Americans in the early twentieth century, see Douglas A. Blackmon's eye-opening book *Slavery by Another Name: The Re-Enslavement of Black Americans from the Civil War to World War II* (New York: Doubleday, 2008).

7. T. J. Curran, *Xenophobia and Immigration, 1820–1930* (Boston: Thwayne, 1975); D. Roediger, *Towards the Abolition of Whiteness: Essays on Race, Politics and Working Class History* (London: Verso, 1994); M. F. Jacobson, *Barbarian Virtues: The United States Encounters Foreign Peoples at Home and Abroad* (New York: Hill and Wang, 2000). Walker's remarks are quoted in Jacobson, 157.

8. A. T. Vaughan, *Roots of American Racism: Essays on the Colonial Experiment* (Oxford: Oxford University Press, 1995), 33.

9. G. B. Nash, *Race and Revolution* (New York: Rowman and Littlefield, 1990), 178.

10. E. Machery and L. Faucher, "Social Construction and the Concept of Race," *Philosophy of Science* 72 (2005): 1208–1219.

11. For example, M. Banton, *The Idea of Race* (Boulder, CO: Westview Publishers, 1978); M. Banton, *The Idea of Race* (Cambridge: Cambridge University Press, 1987); B. Anderson, *Imagined Communities: Reflections on the Origin and Spread of Nationalism* (London: Verso, 1983); W. D. Jordan, *White Over Black: American Attitudes Toward the Negro, 1550–1812* (New York: Norton, 1968); E. Morgan, *American Slavery, American Freedom: The Ordeal of Colonial Virginia* (New York: Norton, 1975); M. Harris, *Patterns of Race in America* (New York: Walker, 1964); A. Smedley, *Race in North America: Origin and Evolution of a Worldview* (Boulder, CO: Westview, 1993).

12. Edouard Machery and Luc Faucher first suggested the label "cognitive-evolutionary approach" in their excellent paper, "Why Do We Think Racially?" published in H. Cohen and C. Lefebvre (eds.), *Handbook of Categorization in Cognitive Science*, eds. H. Cohen and C. Lefebvre (Philadelphia: Elsevier).

13. L. A. Hirschfeld, *Race in the Making: Cognition, Culture and the Child's Construction of Human Kinds* (Cambridge, MA: MIT Press, 1998), xi.

14. A. Nevins (ed.), *George Templeton Strong's Diary of the Civil War, 1860–1865* (New York: Gramercy Books, 1962), 342–43.

15. Plato's comment is from *Phaedrus* 265d–266a.

16. W. Wagner, et al., "An Essentialist Theory of 'Hybrids': From Animal Kinds to Ethnic Categories and Race," *Asian Journal of Social Psychology* (forthcoming).

17. C. W. Mills, *Blackness Visible: Essays on Philosophy and Race* (Ithaca: Cornell University Press, 1998), 60.

18. Ibid., 61. As well as making a philosophical point about race, Mills is also cracking a sly joke. The name *Schwarzenegger* can be construed as "schwarze Neger," which means "black negro," although the true etymology is probably "schwarzen Egger" ("black plowman"), or perhaps "one from Schwarzenegg."

19. C. W. Kalish, "Essentialism to Some Degree: Beliefs About the Structure of Natural Kind Categories," *Memory and Cognition* 30, no. 3 (2002): 340–352. W. Z. Ripley, *Races of Europe*, quoted in I. Hannaford, *Race: The History of an Idea in the West* (Washington, DC: Woodrow Wilson Center Press, 1996), 329.

20. Mills, *Blackness Visible*, 46. Possible-world aficionados will note that I am not assuming a Lewisian notion of worlds, which would forbid transworld identity.

21. L. A. Hirschfeld, "Who Needs a Theory of Mind?," in *Biological and Cultural Bases of Human Inference*, ed. R. Viale, D. Andler, and L. Hirschfeld (Mahwah, NJ: Lawrence Erlbaum Associates, 2006), 154–155. Hirschfeld's citations are to L. Gordon, *The Great Arizona Orphan Abduction* (Cambridge, MA: Harvard University Press, 1999), and R. A. Hahn, J. Mulinare, and S. M. Teutsch, "Inconsistencies in Coding of Race and Ethnicity Between Birth and Death in US Infants: A New Look at Infant Mortality, 1983 Through 1985," *Journal of the American Medical Association* 267 (1992): 259–263.

22. Quoted in I. M. Resnick, "Medieval Roots of the Myth of Jewish Male Menses," *Harvard Theological Review* 93, no. 3 (2000): 259.

23. B. Malamud, *The Fixer* (New York: Farrar, Straus and Giroux, 1966), 139. Quoted in Resnick, 242.

24. D. Sperber, "Pourquois les animaux parfaits, les hybrids et les monstres sont-ils bon à penser symboliquement?," *L'homme* 15 (1975): 22. Cited in English translation in S. Atran, *Cognitive Foundations of Natural History: Towards an Anthropology of Science* (Cambridge: Cambridge University Press, 1990), 59.

25. S. A. Gelman, *The Essential Child: Origins of Essentialism in Everyday Thought* (New York: Oxford University Press, 2003), 10. See also D. Medin, "Concepts and Conceptual Structure," *American Psychologist* 44 (1989): 1469–1481.

26. G. E. Newman and F. C. Keil, "Where Is the Essence? Developmental Shifts

in Children's Beliefs about Internal Features," *Child Development* 79, no. 5 (2008): 1353.

27. M. J. Harner, *The Jívaro: People of the Sacred Waterfall* (Berkeley: University of California Press, 1973), 149.

28. C. S. Brown, *Refusing Racism: White Allies and the Struggle for Civil Rights* (New York: Teachers College Press, 2002), 14.

29. B. Russell, *Unpopular Essays* (New York: Routledge, 1995).

30. A. Rao, "Blood, Milk and Mountains: Marriage Practice and Concepts of Predictability Among the Bakkarwal of Jammu and Kashmir," in *Culture, Creation and Procreation: Concepts of Kinship in South Asian Practice*, eds. M. Böck and A. Rao (New York: Berghahn Books, 2000), 107.

31. J. Golden, *A Social History of Wet-Nursing in America* (New York: Cambridge University Press, 1996), 152–153. Terhune's anecdote is from M. Harland, *Eve's Daughters: Common Sense for Maid, Wife and Mother* (New York: J. R. Anderson, 1882), 30–32. Winters's remarks are from J. E. Winters, "The Relative Influences of Maternal and Wet-nursing on Mother and Child," *Medical Record* 30 (1886), 513. For lactational heredity in seventeenth-century France see C. C. Fairchilds, *Domestic Enemies: Servants and Their Masters in Old Régime France* (Baltimore: Johns Hopkins University Press, 1984) and E. Marvick, "Nature versus Nurture: Patterns and Trends in Seventeenth Century French Childrearing," in *History of Childhood*, ed. L. de Mause (New York: Psychohistory Press, 1974). For Dutch colonists, see A. L. Stoler, *Race and the Education of Desire: Foucault's "History of Sexuality"and the Colonial Order of Things* (Durham, NC: Duke University Press).

32. F. J. Davis, *Who Is Black? One Nation's Definition* (University Park, PA: Pennsylvania State University Press, 1991).

33. W. Wagner et al., "An Essentialist Theory of 'Hybrids': From Animal Kinds to Ethnic Categories and Race," *Asian Journal of Social Psychology*, in press.

34. The term *ethnoraces* was coined by University of California philosopher David Theo Goldberg, and also used by Lawrence Hirschfeld. See D. T. Goldberg, *Racist Culture: Philosophy and the Politics of Meaning* (London: Blackwell, 1993).

35. M. Raudsepp and W. Wagner, "The Essentially Other: Representational Processes That Divide Groups," in *Trust and Distrust Between Groups: Interaction and Representations*, eds. I. Marková et al. (forthcoming).

36. S. Atran, "Folk Biology and the Anthropology of Science: Cognitive Universals and Cultural Particulars," *Behavioral and Brain Sciences* 21 (1998): 547–569.

37. G. Kober, *Biology Without Species: A Solution to the Species Problem* (unpublished Ph.D. dissertation, Boston University, 2009). D. N. Stamos, *The Species*

Problem, Biological Species, Ontology, and the Metaphysics of Biology (Lanham, MD: Lexington Books, 2004).

38. The notion of modularity was introduced into cognitive science by the philosopher Jerry Fodor. Fodor's original concept of cognitive modules is quite different from the notion of Darwinian modules later developed by evolutionary psychologists. For a good account of both Fodorian and Darwinian versions of modularity, see J. L. Bermúdez, *Philosophy of Psychology: A Contemporary Introduction* (New York: Routledge, 2005). An excellent resource for debates about modularity and innateness is the collection of papers in the three volumes of P. Carruthers, S. Laurence, and S. Stich (eds.), *The Innate Mind* (New York: Oxford University Press, 2005, 2006, 2007).

39. P. Boyer and C. Barratt, "Domain Specificity and Intuitive Ontology," in *Handbook of Evolutionary Psychology*, ed. D. M. Buss (New York: Wiley, 2005), 97. Boyer and Barratt cite A. W. Young, D. Hellawell, and D. C. Hay, "Configurational Information in Face Perception," *Perception* 16, no. 6 (1987): 747–759; J. Tanaka and J. A. Sengco, "Features and Their Configuration in Face Recognition," *Memory and Cognition* 25, no. 5 (1997): 583–592; M. Farah, K. D. Wilson, H. M. Drain and J. R. Tanaka, "The Inverted Face Inversion Effect in Prosopagnosia: Evidence for Mandatory, Face-Specific Perceptual Mechanisms," *Vision Research* 35 (1995): 2089–2093; J. Morton and M. Johnson, "CONSPEC and CONLERN: A Two-Process Theory of Infant Face Recognition," *Psychological Review* 98 (1991): 164–181; O. Pascalis, S. de Schonen, J. Morton, C. Druelle et al., "Mothers Face Recognition by Neonates: A Replication and an Extension," *Infant Behavior and Development* 18, no 1 (1995): 79–85; A. Slater and P. C. Quinn, "Face Recognition in the New-born Infant," *Infant and Child Development Special Issue: Face Processing in Infancy and Early Childhood* 10, nos. 1–2 (2001): 21–24; M. Farah, "Specialization Within Visual Object Recognition: Clues From Prosopagnosia and Alexia," in G. R. Martha and J. Farah (eds.) *The Neuropsychology of High-Level Vision: Collected Tutorial Essays. Carnegie Mellon Symposia on Cognition* (Hillsdale, NJ: Lawrence Erlbaum, 1994); B. C. Duchaine, "Developmental Prosopagnosia with Normal Configural Processing," *Neuroreport: For Rapid Communication of Neuroscience Research* 11, no. 1 (2000): 79–83; P. Michelon and I. Biederman, "Less Impairment in Face Imagery Than Face Perception in Early Prosopagnosia," *Neuropsychologia* 31, no. 4 (2003): 421–441; N. Kanwisher, J. McDermott, and M. M. Chun, "The Fusiform Face Area: A Module in Human Extrastriate Cortex Specialized for Face Perception," *Journal of Neuroscience* 17, no. 11 (1997): 4302–4211; J. V. Haxby, E. A. Hoffman, and M. I. Gobbini, "Human Neural Systems for Face Recognition and Social Communication," *Biological Psychiatry* 51, no. 1 (2002): 59–67.

40. For brain damage, see J. R. Anderson, *Cognitive Psychology and Its Implications* (New York: W. H. Freeman and Company, 2005). For the folk-biological module, see S. Atran, "Folk biology and the anthropology of science," *Behavioral and Brain Sciences* 21, no. 4 (1998): 547–569. For critical views see the responses to Atran's target article in *Behavioral and Brain Sciences* 21, no. 4 (1998): 547–569.

41. F. C. Keil, *Concepts, Kinds and Cognitive Development* (Cambridge, MA: MIT Press, 1989), 190, 205.

42. Ibid., 215.

43. Ibid., 176,180.

44. Hirschfeld, *Race in the Making*, 96.

45. Ibid., 98.

46. Ibid., 138. The first anecdote is from P. Ramsay, "Young Children's Thinking about Ethnic Differences," in *Children's Ethnic Socialization*, eds. J. Phinney and M. Rotherham (London: Sage, 1987), 60.

47. For his arguments in favor of this view, and against its most plausible alternative, see L. A. Hirschfeld, "Who Needs a Theory of Mind?" in R. Viale, D. Andler and L. A. Hirschfeld, *Biological and Cultural Bases of Human Inference* (Mahwah, NJ: Lawrence Erlbaum Associates, 2006). For a critical assessment of his research and conclusions, see E. Machery and L. Faucher, "Why Do We Think Racially?" in *Handbook of Categorization in Cognitive Science*, eds. H. Cohen and C. Lefebvre (New York: Elsevier, 2005).

48. K. W. Wamwere, *Negative Ethnicity: From Bias to Genocide* (New York: Seven Stories Press, 2003), 45.

49. See Machery and Foucher, 2005, who cite R. G. Klein, *The Human Career: Human Biological and Cultural Origins* (Chicago: University of Chicago Press, 1999) and, importantly, P. J. Richerson and R. Boyd, "The evolution of human ultra-sociality," in I. Eibl-Eibesfeldt and F. K. Salter, *Indoctrinability, Ideology and Warfare* (New York: Berghahn Books, 1998) and P. J. Richerson and R. Boyd, "Complex Societies: the Evolution of a Crude Superorganism," *Human Nature* 10: 253–289.

50. *Judges* 12: 5–6.

51. E. H. Erikson, "Pseudospeciation in the Nuclear Age," *Political Psychology*, 6, no. 2 (1986): 214.

52. F. J. Gil-White, "Are Ethnic Groups Biological 'Species' to the Human Brain?," *Current Anthropology* 42, no. 4: 519.

53. Ibid., 432.

7. THE CRUEL ANIMAL

1. D. Penn, K. J. Holyoak and D. Povinelli, "Darwin's Mistake: Explaining the Discontinuity Between Human and Nonhuman Minds," *Behavioral and Brain Sciences*, 31(2008), 109.

2. W. Shakespeare, *King Lear*, I. iv.

3. M. Twain, *What Is Man?: And Other Philosophical Writings* (Berkeley: University of California Press, 1973), 84–85.

4. E. O. Wilson and B. Hölldobler, *Journey to the Ants: A Story of Scientific Exploration* (Cambridge, MA: Harvard University Press, 1994), 59.

5. R. Wrangham and D. Peterson, *Demonic Males: Apes and the Origin of Human Violence* (New York: Matiner, 1997), 219.

6. Ibid., 70.

7. J. Goodall, *Through a Window: My Thirty Years with the Chimpanzees of Gombe* (New York: Houghton Mifflin, 1990), 108–109.

8. R. W. Wrangham, "Evolution of Coalitionary Killing," *Yearbook of Physical Anthropology* 42, no. 1(1999): 1–30.

9. Ibid., 22.

10. R. B. Ferguson, "Materialist, Cultural and Biological Theories on Why Yanomami Make War," *Anthropological Theory* 1, no. 1 (2001): 106.

11. Wrangham and Peterson, *Demonic Males*, 14.

12. Chagnon, 128–130.

13. Wrangham and Peterson, *Demonic Males*, 18.

14. Goodall, *Through a Window*, 210.

15. B. Hölldobler and E. O. Wilson, *The Ants* (Cambridge, MA: Harvard University Press, 1990).

16. Anderson developed the concept of imagined communities to explain nationalism, but he recognized that it has a much wider application, writing that "all communities larger than primordial villages of face-to-face contact (and perhaps even these) are imagined." B. Anderson, *Imagined Communities* (New York: Verso, 1983), 6.

17. K. Vonnegut, *Galápagos: A Novel* (New York: Dial Press, 1999), 9.

18. C. Darwin, *The Descent of Man and Selection in Relation to Sex* (Charleston, SC: Bibliolife, 2009), 168.

19. D. Hume, *Dialogs and Natural History of Religion* (New York: Oxford University Press, 2009), 141.

20. S. Guthrie, *Faces in the Clouds: A New Theory of Religion* (New York: Oxford University Press, 1993), 62, 91.

21. Goodall, *Through a Window*, 108–109.

22. See D. Bickerton, *Adam's Tongue: How Humans Made Language, How Language Made Humans* (New York: Hill and Wang, 2009).

23. W. James, *The Principles of Psychology* (Cambridge, MA: Harvard University Press, 1981), 462.

24. D. J. Goldhagen, *Worse Than War: Genocide, Eliminationism, and the Ongoing Assault on Humanity* (New York: PublicAffairs), 191–192.

25. G. Berkeley, "Three Dialogs between Hylas and Philonous," in *Principles of Knowledge/Three Dialogues* (London: Penguin, 1988), 195.

8. AMBIVALENCE AND TRANSGRESSION

1. S. Freud, "Totem and Taboo," in *The Complete Psychological Works of Sigmund Freud*, trans. J. Strachey, vol. 13 (London: Hogarth Press and the Institute for Psycho-Analysis, 1950), 65.

2. D. A. Grossman, *On Killing: The Psychological Cost of Learning to Kill in War and Society* (Boston: Little, Brown & Co., 1996), 92–93.

3. G. L. Vistica, "One Awful Night in Tanh Phong," *New York Times Magazine*, April 29, 2001, 131.

4. B. Shalit, *The Psychology of Combat and Conflict* (New York: Praeger, 1988), 2. Cited in D. Grossman, *On Killing*.

5. J. G. Gray, *The Warriors: Reflections on Men in Battle* (Lincoln: University of Nebraska Press, 1998), 51.

6. E. Jünger, *Der Kampf als inneres Erlebnis* (Berlin: E.G. Mittler & Sohn, 1925). Quoted in translation in Gray, *The Warriors*, 52.

7. W. Broyles Jr., "Why Men Love War," *Esquire*, November 1984, 57.

8. J. Ramirez, "Carnage.com," *Newsweek*, May 10, 2010. The term *war porn* comes from the French philosopher Jean Baudrillard's paper of the same name, in *The Conspiracy of Art: Manifestos, Texts, Interviews* (Cambridge, MA: MIT Press, 2005).

9. S. L. A. Marshall, *Men Against Fire: The Problem of Battle Command* (Norman, OK: University of Oklahoma Press, 2000), 78–79. In recent years Marshall's claims have come under fire. See R. J. Spiller, "S. L. A. Marshall and the Ratio of Fire," *RUSI* Journal (1988): 63–71. E. Thomas, "Fire Away: Exploding One of Military History's More Enduring Myths," *Newsweek*, December 12, 2007. For a more balanced assessment, see K. C. Jordan, "Right for the Wrong Reasons: S. L. A. Marshall and the Ratio of Fire in Korea," *Journal of Military History* 66, no. 1(2002): 135–162

10. R. A. Kulka et al., *Trauma and the Vietnam War Generation: Report of Findings from the National Vietnam Veterans Readjustment Study* (New York: Brunner/ Mazel, 1990).

11. Marshall, *Men Against Fire*, 78.

12. W. Manchester, *Goodby Darkness: A Memoir of the Pacific War* (New York: Dell, 1980), 17–18.

13. In E. C. Johnson (ed.), *Jane Adams: A Centennial Reader* (New York: Macmillan, 1960), 273.

14. R. M. MacNair, *Perpetration-Induced Traumatic Stress: The Psychological Consequences of Killing* (New York: Praeger/Greenwood, 2002), 47.

15. A. Fontana and R. Rosenheck, "A Model of War Zone Stressors and Posttraumatic Stress Disorder," *Journal of Traumatic Stress* 12 (1999): 111–126; R. M. MacNair, "Perpetration-inducted Traumatic Stress in Combat Veterans," *Peace and Conflict: Journal of Peace Psychology* 8 (2002): 63–72; S. Maguen *et. al.*, "The Impact of Killing in War on Mental Health Symptoms and Related Functioning," *Journal of Traumatic Stress* 22, no. 5 (2009): 435–443. M. S. Kaplan, et al. "Suicide Among Male Veterans: A Prospective Population-based Study, *Journal of Epidemiology and Community Health* 61 (2007): 619–624. H. Hendon and A. P. Haas, "Suicide and Guilt as Manifestations of PTSD in Vietnam," *Journal of American Psychiatry* 138 (1991): 586–591. http://www.ptsd.va.gov/public/pages/ ptsd-suicide.asp>. See also J. E. S. Phillips, *None of Us Were Like This Before: American Soldiers and Torture* (New York: Verso, 2010).

16. J. Shay, *Odysseus in America: Combat Trauma and the Trials of Homecoming* (New York: Scribner, 2002); B. T. Litz et al., "Moral Injury and Moral Repair in War Veterans: A Preliminary Model and Intervention Strategy," *Clinical Psychology Review* 29 (2009), 700.

17. Litz et al.; B. P. Marx, "Posttraumatic Stress Disorder and Operations Enduring Freedom and Iraqi Freedom: Progress in a Time of Controversy," *Clinical Psychology Review* 29 (2009): 671–673.

18. Ibid., 697.

19. J. Diamond, "Vengeance Is Ours: Annals of Anthropology," *The New Yorker*, April 21, 2008, 74.

20. Diamond's article eventuated in a ten million dollar law suit by his informant. C. Silverman, "New Yorker under Siege," *Columbia Journalism Review*, May 22, 2009.

21. G. Orwell, "Looking Back on the Spanish War," *Facing Unpleasant Facts: Narrative Essays* (New York: Houghton Mifflin Harcourt, 2008), 149.

22. E. Lussu, *Sardinian Brigade* (New York: Alfred A. Knopf, 1939), 174.

23. Dean, 108.

24. Eibl-Eibesfeldt, *Human Ethology* (New York: Aldine de Gruyter, 1989), 420.

25. H. H. Turney-High, *Primitive War: Its Practices and Concepts* (Charleston, SC: University of South Carolina Press, 1991), 225. See also J. M. G. van der Dennen, "The Politics of Peace in Primitive Societies: The Adaptive Rationale Behind Corroboree and Calumet," in I. Eibl-Eibesfeldt and F. K. Salter, *Ethnic Conflict and Indoctrination: Altruism and Identity in Evolutionary Perspective* (New York: Berghahn Books, 1998).

26. R. Sapolsky, "A Natural History of Peace," *Foreign Affairs* 85, no. 1 (2006): 119.

27. R. Wrangham, *Catching Fire: How Cooking Made Us Human* (New York: Basic Books, 2009).

28. D. L. Cheney and R. M. Seyfarth, *How Monkeys See the World: Inside the Mind of Another Species* (Chicago: University of Chicago Press, 1992).

29. D. G. Bates and J. Tucker (eds.), *Human Ecology: Contemporary Research and Practice* (New York: Springer, 2010).

30. A. Clark, *Supersizing the Mind: Embodiment, Action, and Cognitive Extension* (New York: Oxford University Press, 2008), 58.

31. M. Kawai, "Newly-acquired Pre-cultural Behavior of the Natural Troop of Japanese Monkeys on Koshima Islet," *Primates* 6, no. 1 (1965): 1–30.

32. D. Dennett, *Consciousness Explained* (New York: Little Brown and Company, 1991), 195.

33. S. J. Mithen, *The Prehistory of The Mind: The Cognitive Origins of Art, Religion, and Science* (London: Thames and Hudson, 1999); S. J. Mithen, *The Singing Neanderthals: The Origins of Music, Language, Mind and Body* (Cambridge, MA: Harvard University Press, 2007); S. J. Mithen and L. Parsons, "The Brain as a Cultural Artifact," *Cambridge Archaeological Journal* 18, no. 3. (2008): 415–422; S. Mithen, "Ethnobiology and the Evolution of the Human Mind," *Journal of the Royal Anthropological Society* 12, no. 1 (2005): 45–61. See also P. Carruthers, "The Cognitive Functions of Language," *Behavioral and Brain Sciences*, 25 (2002): 657–726.

34. H. Ofek, *Economic Origins of Human Evolution* (New York: Cambridge University Press, 2001), 172–173.

35. O. Bar-Yosef, "The Archaeological Framework of the Upper Paleolithic Revolution," *Diogenes* 54 (2007): 3–18

36. S. L. Kuhn and M. C. Stiner, "Paleolithic Ornaments: Implications for Cognition, Demography and Identity," *Diogenes* 54 (2007), 47–48.

37. J. Gulaine and J. Zammit, *The Origins of War: Violence in Prehistory* (Malden, MA: Blackwell, 2005), 55–56.

38. C. Spencer, *British Food: An Extraordinary Thousand Years of History* (New York: Columbia University Press, 2002), 149.

39. A. Hyatt Verrill, *Wonder Creatures of the Sea* (New York: D. Appleton-Century Company, 1940).

40. Philipians, 3:2; "Iranian Cleric Denounces Dog Owners," BBC News, October 14, 2002 <http://news.bbc.co.uk/2/hi/middle_east/2326357.stm>

41. N. Z. Davis, "Religious Riot in Sixteenth-Century France," in *The Massacre of St. Bartholomew: Reappraisals and Documents*, ed. A. Soman (The Hague: Martinus Nijhoff, 1974), 209–219.

42. Gray, *The Warriors*, 148.

43. Rick Atkinson and Dan Boltz, "Allied Bombers Strike Shifting Iraqi Troops," *The Independent*, February 6, 1991; "The Reality of War," *Observer*, March 30, 2003; "Perspectives," *Newsweek*, May 3, 2004.

44. D. Quammen, *Monster of God: The Man-Eating Predator in the Jungles of History and the Mind* (New York: W. W. Norton and Company, 2003), 3.

45. B. Ehrenreich, *Blood Rites: Origins and History of the Passions of War* (New York: Henry Holt and Company, 1997), 49.

46. E. Gardiner, *Visions of Heaven and Hell Before Dante* (New York: Italica, 1989), 140.

47. J. E. Salisbury, *The Beast Within: Animals in the Middle Ages* (New York: Routledge, 1994). E. Gardiner, *Visions of Heaven and Hell Before Dante* (New York: Italica Press, 2008). Augustine, "The Christian Combat," in *The Writings of Saint Augustine*, vol. 4 (New York: Fathers of the Church, 1947), 317.

48. M. Perry and F. M. Schweitzer, *Antisemitism: Myth and Hate from Antiquity to the Present* (New York: Palgrave Macmillan, 2002), 76.

49. P. Krentz and E. L. Wheeler (eds. and trans.), *Polyaenus: Stratagems of War II* (Chicago: Ares, 1994), 625.

50. These quotations are from A. Malouf, *The Crusades Through Arab Eyes*, trans. J. Rothschild (New York: Schocken Books, 1984), 39.

51. S. Keen, *Faces of the Enemy: Reflections on the Hostile Imagination* (New York: HarperCollins, 1986).

52. *Iliad*, 15: 604–30, 24: 207, 24: 43, 16: 155–67, quoted in Gottschall, 280.

53. B. Ehrenreich, *Blood Rites*, 11.

54. A. Palmer, *Colonial Genocide* (Adelaide: C. Hurst & Co. Ltd, 2001), 44f; J. La Nause, *The Making of the Australian Constitution* (Melbourne: Melbourne University Press, 1972); J. La Nause, *The Making of the Australian Constitution* (Melbourne: Melbourne University Press, 1972); "'Plenty Shoot 'Em': The Destruction of Aboriginal Societies Along the Queensland Frontier," in *Genocide and Settler Society: Frontier Violence and Stolen Indigenous Children in Australian History*, ed. A. D. Moses (New York: Berghahn Books, 2004), 155, 159.

55. Herodotas, *Histories*, trans. G. Rawlinson (London: Wordsworth, 1999), 326.

56. R. J. Chacon and D. H. Dye (eds.), *The Taking and Displaying of Body Parts as Trophies by Amerindians* (New York: Springer, 2007), 7.

57. R. F. Murphy, "Intergroup Hostility and Social Cohesion," *American Anthropologist* 59 (1957), 1025–1026, 1028.

58. J. R. Ebert, *A Life in a Year: The American Infantryman in Vietnam, 1965–1972* (Novato, CA: Presidio), 280.

59. P. Fussell, "Thank God for the Atomic Bomb," in *Thank God for the Atomic Bomb and Other Essays* (New York: Summit, 1988), 26.

60. P. Fussell, *Wartime: Understanding and Behavior in the Second World War* (Oxford: Oxford University Press, 1989), 117.

61. P. Fussell, "Postscript (1987) on Japanese Skulls," in *Thank God for the Atomic Bomb and Other Essays* (New York: Summit, 1988), 48.

62. A. Santoli, *Everything We Had: An Oral History of the Vietnam War by Thirty-Three American Soldiers Who Fought It* (New York: Ballantyne, 1981), 98–99.

9. QUESTIONS FOR A THEORY OF DEHUMANIZATION

1. S. Freud, *Civilization and Its Discontents* (New York: W.W. Norton and Company, 1961), 97.

2. P. Watson, *War on the Mind: The Military Uses and Abuses of Psychology* (New York: Basic Books, 1978), 250.

3. R. Rorty, "Human Rights, Rationality and Sentimentality," in *Truth and Progress* (New York: Cambridge University Press, 1998).

4. D. Rieff, "Letter from Bosnia," *New Yorker*, November 23, 1992, 82–95.

5. Rorty, "Human Rights, Rationality and Sentimentality," 167.

6. Ibid., 168.

7. Ibid., 178.

8. Ibid., 176, 185.

9. Cited in A. Alvarez, *Governments, Citizens, and Genocide: A Comparative and Interdisciplinary Approach* (Indianapolis: Indiana University Press, 2001), 151.

10. Rorty, "Human Rights, Rationality and Sentimentality," 180.

11. Ibid., 169–170.

APPENDIX II: PAUL ROSCOE'S THEORY OF DEHUMANIZATION IN WAR

1. P. Roscoe, "Intelligence, Coalitional Killing, and the Antecedents of War," *American Anthropologist* 109, no. 3 (2007): 485–495.

2. C. Browning, *Ordinary Men: Reserve Police Battalion 101 and the Final Solution in Poland* (New York: HarperCollins, 1998).

3. Roscoe, "Intelligence, Coalitional Killing, and the Antecedents of War," 487.

4. Ibid., 488.

5. Ibid., 490.

INDEX

MAR 2 9 2011